WEST COAST
FOSSILS

WEST COAST
FOSSILS

A Guide to the Ancient Life of Vancouver Island

Second Edition

ROLF LUDVIGSEN
AND
GRAHAM BEARD

HARBOUR PUBLISHING

Second printing, 2001

Harbour Publishing
P.O. Box 219
Madeira Park, BC
V0N 2H0
Canada

Cover, page design, and composition by Martin Nichols, Lionheart Graphics
Cover photograph by Paul Bailey
Interior photographs by Rolf Ludvigsen and Graham Beard, except for Figures 21, 24, 27, 28 and 44, which were supplied by the Geological Survey of Canada, Figures 124 and 131 by K. Morrison and B. Hessin, and Figure 142 by M. McColl
Edited by Maggie Paquet
Maps by Joe Morin

We acknowledge the financial support of the Government of Canada through the Book Publishing Industry Development Program for our publishing activities.

Printed and bound in Canada

Canadian Cataloguing in Publication Data

Ludvigsen, Rolf, 1944-
West Coast fossils

Includes bibliographical references and index.
ISBN 1-55017-179-8

1. Fossils—British Columbia—Vancouver Island. 2. Fossils—Collection and preservation. I. Beard, Graham, 1941- II. Title.
QE705.C3L83 1998 560'.9711'2 C98-910222-X

Contents

Acknowledgements

We have been helped by many paleontologists, curators, and private collectors who gave us access to their own fossil collections or who shared information about the identity and significance of Vancouver Island fossils.

Most of the specimens photographed for the book were collected by Graham and Tina Beard over a twenty-year period. These fossils are now in the collections of the Vancouver Island Paleontological Museum (VIPM) in Qualicum Beach. For allowing us to photograph other specimens, we thank Tom Bolton and Terry Poulton of the Geological Survey of Canada (GSC) in Ottawa and Calgary, Richard Hebda of the Royal British Columbia Museum (RBCM) in Victoria, and Deborah Griffiths-Ferguson of the Courtenay and District Museum (CDM). We are also grateful to Peter Bock of Nanaimo, Dan Bowen of Courtenay, Bob Copeman of Victoria, Joe Haegert of Victoria, Beryl Geppert of Hornby Island, Bill Irwin of Port Alberni, Tim Obear of Courtenay, and Mike Trask of Courtenay for allowing us to photograph fossil specimens from their private collections; many of these specimens have been donated to different Vancouver Island museums.

Many paleontologists shared their specialized knowledge of Vancouver Island fossils with us. In particular, we acknowledge the help of Jim Haggart of the Geological Survey of Canada, Vancouver (Cretaceous mollusks); J.D. Stewart of the Los Angeles County Museum (Cretaceous fishes); Paul Smith of the University of British Columbia, Vancouver (Jurassic ammonites); Betsy Nicholls of the Royal Tyrrell Museum of Paleontology, Drumheller (Cretaceous reptiles); Lee McKenzie McAnally of the University of Victoria (Cenozoic mammals); and Andrew Ross of the Booth Museum in Brighton, England (Cretaceous insects).

Several individuals gave generously of both their time and their knowledge to help us frame important parts of this book. Tim Tozer of the Geological Survey of Canada in Vancouver helped us with Triassic fossils and geology of Vancouver Island and provided us with photographs, some previously unpublished. Jim Basinger and Beth McIver of the University of Saskatchewan in Saskatoon responded at length and with great insight to our inquiries about the identity and meaning of the Vancouver Island Cretaceous plants. Joe Morin of Courtenay undertook the time-consuming task of drafting all of the geological maps of Vancouver Island. Louise Bell of Denman Island made numerous editorial suggestions that greatly improved the clarity of presentation.

The cost of preparing the photographic illustrations was covered by a generous grant from the Canadian Geological Foundation—a worthy private agency that supports projects promoting public interest in the geological sciences.

Finally, Tina Beard has been a not-so-silent partner in this project for the past few years. She drew the striking sea-bottom dioramas and, in general, helped move the work along when it stalled, as it did many times.

Preface to the Second Edition

The last decade has seen a remarkable upsurge of interest in local fossils on Vancouver Island. We like to think that *West Coast Fossils* has played a part. Even though fewer professional paleontologists work among the rocks of the island, more—and different—fossils are now being discovered by increasing numbers of fossil collectors and paleontologists. Most of these, amateur as well as professional, are members of the British Columbia Paleontological Alliance, an umbrella group of individuals and organizations dedicated to increasing awareness and understanding of fossils in the Cordillera. Amateur paleontologists have made virtually all of the recent discoveries of new and exciting fossils on Vancouver Island. Consider the following items:

- In a single-minded campaign to find the first trilobites on Vancouver Island, a paleontologist systematically surveys the muddy limestones exposed on logging roads east of Alberni Inlet. Eventually, he discovers them at Rift Creek along with a rich fossil fauna of Late Carboniferous age.

- A then-novice fossil collector, who didn't know that it was pointless to look for fossils among the volcanic rocks south of Duncan, discovers the first Middle Jurassic fossils known on Vancouver Island in a siltstone pocket along the Koksilah River.
- A bulldozer operator moving blasted rock along the approach to the new ferry terminal at Duke Point notes a peculiar pattern in a large block and sets it aside. Later, fossil collectors discover an important Upper Cretaceous fossil flora.
- A couple of amateur paleontologists begin to mine a thin Upper Cretaceous shale unit on Hornby Island for fossil shark teeth. They find more than a dozen species and, in the process, discover the first pterosaur and beetle fossils known for the west coast.
- A curator of a North Island museum stops to split Upper Triassic shales where the Island Highway crosses the Keogh River and discovers the first articulated fossil fish on Vancouver Island.
- A group of amateur paleontologists begin to collect fossils from supposed Upper Cretaceous rocks south of Campbell River and discover the first marine Cenozoic rocks on the east coast of Vancouver Island, as well as exceptionally preserved fossil seeds and nuts.
- An observant amateur notes that blasting to widen a logging road along the Memekay River south of Sayward has exposed a large siltstone pocket containing the best Lower Jurassic fossils yet discovered on Vancouver Island.
- An amateur paleontologist and his son find a peculiar tooth in an Upper Cretaceous shale block on the Trent River. The fossil tooth turns out to be the first evidence that dinosaurs lived west of the Rocky Mountains.

All of these Vancouver Island sites and fossils were discovered after the manuscript for the first edition of *West Coast Fossils* was completed in fall 1993. In order to incorporate these fossils in this expanded second edition, we have prepared twenty-seven new figures to adequately show these additional fossils.

This new edition would not have been possible without the help of our friends and colleagues who allowed us to study and photograph the Vancouver Island fossils in their collections. In addition to those named in the Acknowledgements above, we thank the following: Bill Hessin of Qualicum Beach for lending specimens and providing information about the Late Carboniferous fossils he discovered at Rift Creek, and for allowing us to use his trilobite reconstructions; Kurt Morrison of Hornby Island for information about the fossil shark teeth and other fossils he collected at Collishaw Point, and for allowing us to use his and Bill's photographs of the teeth and the pterosaur bone; Jack Whittles of Ladysmith for giving us access to his large collection of Cretaceous plants from the Cranberry Arms site and fossil brittle stars from Maple Bay; Chris Ruttan of Shawnigan Lake for lending us his collection of Middle Jurassic fossils from the Koksilah River and the unique crinoid from Haslam Creek; William Reeve of the Port Hardy Museum for lending us the articulated Triassic fish he collected at the Keogh River; and Mike Orchard and Tim Tozer of the Geological Survey of Canada in Vancouver for sending us photographs of conodonts and ammonites, respectively. We also thank Conrad Labandeira of the Smithsonian Institution in Washington, DC for identifying the fossil weevil; Terry Poulton of the Geological Survey of Canada in Calgary and LouElla Saul of the Los Angeles County Museum for information about Jurassic and Cretaceous fossils, respectively; and Thor Henrich of Victoria for information about the Cranberry Arms site. Finally, for encouraging us to prepare this second edition and for facilitating the change of publisher, we thank Howard White.

Introduction

There's little doubt that people are fascinated by fossils. Perhaps it is the paradox of once-living stone, perhaps it is the immense age or unusual shapes of these natural objects, or perhaps it is because fossils are unambiguous records of strange life forms of the deep past.

Contrary to popular belief, fossils are not rare. Large, complete, and well-preserved fossils are definitely uncommon, but small fossil shells, bones, teeth, and plant remains occur widely in sedimentary rocks exposed across Canada. However, here on the west coast, most residents and visitors are unaware that important and beautiful fossils are preserved in the sedimentary rocks that lie literally beneath our feet. These rocks are exposed on mountainsides, cliffs, river bottoms, rocky shores, road cuts, and quarries across Vancouver Island and on the Gulf Islands.

Everyone here knows that Vancouver Island is a singular place. The climate is like no other region of Canada, the plants are different, and even the air and water have unique qualities. So it should not come as a surprise that the geological foundations of the island, the rocks that support everything else, and the fossils found in these rocks, are also distinct. Vancouver Island fossils are very similar to those from lands on the far side of the Pacific Ocean or from countries now washed by the Indian Ocean. They differ markedly from fossils found in the rest of Canada. The well-preserved ammonites from Hornby Island, for example, can be matched, almost species for species, with Late Cretaceous ammonites from Pondicherry in southern India, but they are entirely different from ammonites of the same age in nearby Alberta.

Over the past 150 years, fossils have been collected at numerous localities on Vancouver Island and the Gulf Islands, but they still remain poorly known. Most of these fossils have been described and illustrated by paleontologists in monographs, government series, and specialist journals. However, few of these publications are readily available in public libraries or even in college or university libraries on Vancouver Island. Thousands of fossil specimens, rare and commonplace, are kept in collections at federal and provincial geological surveys, at museums, and at universities. Only a few are on public display. In addition, uncounted fossils, including some of the best specimens known, are hidden from public view in private collections.

A wide range of books on fossils and their meaning is available in libraries and bookstores. Some deal with broad themes, such as living fossils or mass extinctions. Others focus on the paleobiology, paleoecology, and evolution of various animal groups—ammonites, trilobites or, the perennial favourites, dinosaurs. Some books attempt the impossible—for example, guidebooks that purport to cover fossils of all ages on all continents. To date, no book has dealt solely with those fossils to be found in the rocks of coastal British Columbia.

By focusing on the fossils of a single region of limited extent, we can be detailed and comprehensive in our coverage. We treat here most of the fossils that a meticulous collector can reasonably expect to find in the rocks of Vancouver Island. Throughout, we have tried to describe interesting features and aspects of each fossil and tell the reader something of the lives of these ancient animals or plants. This information is complemented by photographs of the best Vancouver Island fossil specimens we could find. It is our hope that these illustrations will encourage amateur paleontologists to identify the fossils they have collected. Some of the illustrated specimens were retrieved from dusty museum drawers, others were borrowed from geological survey collections, and still others are from our own collections or begged and borrowed from our friends.

Each fossil in the book is identified by a unique catalogue number, noted in brackets. The prefix indicates the museum in which the specimen is located. Most of the specimens are kept at the Vancouver Island Paleontological Museum (VIPM) in Qualicum Beach. Others are kept at the Geological Survey of Canada (GSC) in Ottawa, the Royal British Columbia Museum (RBCM) in Victoria, or at the Courtenay and District Museum (CDM). Those fossils kept in the private collections of Peter Bock of Nanaimo, Jack Whittles of Ladysmith, Bill Hessin of Qualicum Beach, Kurt Morrison of Hornby Island, Joe Haegert of Victoria, and Beryl Geppert of Hornby Island are identified by the name of the collector only.

To the uninitiated, a description of how fossils are collected reads like punishment, not pleasure. After all, a mental image of people using sledgehammers and chisels to break open hard rocks is more consistent with a chain gang sweating in a quarry than with a group of naturalists enthusiastically searching for records of ancient life. And yet, growing numbers of British Columbians are discovering that the quest for fossils can be a highly satisfying outdoor activity.

Paleontology is often confused with archeology. Even though they share the same historical focus, they are quite different disciplines. An archeologist excavates human remains, artifacts, and dwellings to shed light on the cultural history of early societies. A paleontologist collects and studies naturally occurring fossil remains in order to document the ecological and evolutionary history of animals and plants. The time scale of the archeologist is measured in thousands of years; that of the paleontologist reaches back over hundreds of millions of years.

When the final stone was removed from the entrance to Tutankhamen's tomb on November 29, 1922, Howard Carter was the first person in over three thousand years to peer into that crypt. We can't be part of such archeological excavations; we can only participate vicariously in Carter's excitement as the light from his candle played across the tumble of magnificent artifacts. But every time we break a rock in search of fossils, we may experience a comparable thrill of discovery. When a concretion or a piece of limestone is cracked, or a slab of shale is split, that fracture exposes an ancient surface to the light of day for the first time in many millions of years. And the paleontologist wielding the hammer is the first person ever to examine that scene.

In the course of a single morning, a dedicated fossil collector on one outcrop may examine hundreds of pieces of rock. Most will be passed over, but those that look promising will be cracked open. In most cases, the rock contains nothing of interest. But occasionally, the collector will experience a *frisson* of excitement as the fracture opens to reveal the perfect ammonite—a wondrous and natural icon of a former age.

CHAPTER ONE

Fossils: Keys to Earth History

The Latin word *fossilis* simply means something dug out of the ground—stones, minerals, crystals, potsherds, flint implements, and what used to be called "figured stones." Now "fossil" refers only to the remains of ancient animals and plants naturally preserved in sedimentary rocks. These remains are the essentials of paleontology—a scientific discipline that, in straddling geology and biology, is concerned with the nature and significance of ancient life as represented by fossils preserved in rock.

To paleontologists and geologists, fossils are of inestimable value in deciphering the history of Earth. They provide the best, the easiest, and, arguably, the most accurate method of dating rocks, of measuring geologic time, and of correlating rock strata. Fossils give critical information about past environments and about ancient animal and plant communities. And, of course, they comprise the objective data necessary for determining the course of evolution. In addition, for both amateur and scientist, fossils hold fascination as natural objects of great beauty.

The rocks of Vancouver Island yield a great variety of fossils—delicate insect wings, commonplace clams and crabs, massive reptile vertebrae, razor-sharp shark teeth, blackened sprigs of leaves, and iridescent coiled ammonites. Fossils also include the castings, trails, tracks, tunnels, and excrement left behind by long-vanished animals. Individual fossil specimens range in size from a coiled ammonite the size of a truck tire to a microfossil smaller than a grain of sand. Some fossil bones must have belonged to reptiles 10 m or more in length.

Every fossil was, at one time, an animal or plant that lived and died as part of an ancient ecosystem. If carefully collected, prepared, and identified, each recovered fossil can become a link to life in the deep past—a means by which a naturalist, professional or amateur, can reach back across the eons to touch these inhabitants of past worlds.

Figure 1 Body fossils comprise the actual shell or skeleton of organisms. Three examples are shown: the Cretaceous ammonite *Canadoceras* and a shark tooth (VIPM 001), the Cenozoic crab *Zanthopsis* (CDM), and a Cretaceous reptile bone preserved in a concretion (VIPM 002).

Note: The bar in each figure represents 1 cm.

TYPES OF FOSSILS

The most common and most familiar category of fossil is a body fossil—the actual remains of an organism. In most cases, a body fossil consists of the internal or external skeleton of an animal—shells, bones, teeth, or scales (Figure 1). Such skeletons are readily preserved in sedimentary rock because they are already mineralized. The most common minerals are calcium carbonate (shells of bivalves, gastropods, ammonites, corals, brachiopods, and crustaceans) and calcium phosphate (bones and teeth of vertebrates). Less common is chitin, an organic compound that forms the exoskeleton of insects, and silicon dioxide, a glassy mineral found in some sponges. Plant fossils are preserved as carbon compressions of leaves and needles, and as permineralized wood and cones in which the spaces between the cell walls have been impregnated with a mineral, commonly silica or calcium carbonate.

Figure 2 Trace fossils preserve evidence of the activity of animals. Here, two kinds of Cretaceous sediment-feeding animals left castings of different sizes (VIPM 003).

The other major category of fossil is a trace fossil. Whereas a body fossil gives information about the morphology of an ancient plant or animal, which leads to proper identification, a trace fossil indicates the activity or behaviour of a once-living animal (Figure 2). Trace fossils include footprints, burrows, borings, and bite marks. Fossil feces, or coprolites, also belong in this category. In some cases, the identity of the trace maker is obvious; in others, it cannot be determined.

Figure 3 Pseudofossils are inorganically produced stony objects that mimic real fossils. The "bone" on the left is actually a water-worn pebble of volcanic rock. The lumpy concretions were formed within shale. The "Turtle rock" on the right is a concretion with polygonal cracks on the surface.

Any listing of categories of fossils would be incomplete without mentioning pseudofossils—inorganically produced stony objects that mimic real animals and plants (Figure 3). Some pseudofossils consist of pebbles and rocks that have been sculpted by water and wind into shapes that evoke parts of real animals. The irregularly shaped calcareous concretions that abound in Upper Cretaceous shales of Vancouver Island and the Gulf Islands are, with a bit of imagination, strongly reminiscent of animals and are frequently mistaken for real fossils. The curious "turtle rocks" with regular polygonal cracks on the surface are actually concretions that formed within shales.

IDENTIFYING AND CLASSIFYING FOSSILS

Most marine animal species now living in the Strait of Georgia have common names, such as the sunflower sea star or the purple shore crab. Common names are important because they are descriptive and easily remembered, but they have little meaning, for example, to a Chinese or Russian marine biologist. In order for biologists to discuss and record their findings without confusion, each living animal and plant species also bears a unique scientific name in addition to one or more common names. So, to marine biologists of all countries, the sunflower sea star is known as *Pycnopodia heliathoides* and the purple shore crab as *Hemigraptus nudus*.

This binomial system was devised by the great eighteenth-century Swedish naturalist, Carl Linnaeus. In this system, the scientific name consists of a genus name and a species name, both based on either Latin or Greek roots. By convention, the genus and species names are italicized in print.

People regularly assign living things to various general categories, such as deer, whelks, or firs. Biologists and paleontologists, however, formally classify living and fossil organisms into ever-narrower (or more specific) hierarchical categories starting with kingdom and ending with species. Examples of the formal classification of two fossils, an ammonite and a crab, respectively, are shown below.

KINGDOM	Metazoa	Metazoa
PHYLUM	Mollusca	Arthropoda
CLASS	Cephalopoda	Crustacea
ORDER	Ammonitida	Decapoda
FAMILY	Turrilitidae	Raninidae
GENUS	*Nostoceras*	*Cretacoranina*
SPECIES	*N. hornbyense*	*C. harveyi*

When it comes to identification, it is the detail, not the generality, that counts. No self-respecting bird watcher would be content with recording a "brown sparrow-sized bird." The pursuit continues with binoculars and field guides until identification of the Lapland longspur can be verified and that is the name entered in the log book. Similarly, whether or not a mushroom hunter can distinguish the delicious meadow mushroom from the deadly and superficially similar destroying angel has more than theoretical significance.

The same principles apply to the identification of fossils. A fossil collector may have trouble distinguishing a poorly preserved fossil from a pseudofossil. But a moderately complete fossil from, say, Cretaceous rock of Vancouver Island, is easily recognized as being a bivalve, an ammonite, or a snail. A good amateur paleontologist will view "an ammonite" as the starting point of the process of identification, not the end point. Once a few characteristic features are determined, this fossil can then be assigned to one of a few dozen genera of Cretaceous ammonites—for example, to the genus *Canadoceras*—and, if it is well preserved, to a species—for example, *Canadoceras newberryanum*. Only when the species is determined can the ammonite be considered identified.

Lacking common names, fossil species are identified only by their scientific names. Some of these names are quite a mouthful—for example, the ammonites *Anagaudryceras politissimum* and *Pseudoschloenbachia umbulazi*, and the crab *Longusorbis cuniculosus*. But these names are not without meaning; some are neatly descriptive. For example, the name of the rock-boring date mussel *Lithophaga* is Greek for "stone eater"; the long, slender, and ribbed bivalve *Pinna* is Latin for "feather"; the robust coiled ammonite *Pachydiscus* is Greek for "thick circular plate"; and the curled heteromorph ammonite *Bostrychoceras* is Greek for "ringlet horn."

Poorly preserved or incomplete fossils cannot be identified to the species level and, therefore, are assigned names that reflect such uncertainty. For example, *Solemya* sp. is an undetermined species of the bivalve genus *Solemya*; the ammonite *Tollia* cf. *simplex* is a species of the genus *Tollia* similar to the species *T. simplex* (cf. stands for "confer," Latin for "compare to").

ILLUSTRATING FOSSILS

A fossil may be photographed in its natural state or it may be photographed after it has been prepared so as to highlight its surface sculpture and shape. These two techniques produce strikingly different results, which are compared in the pairs of photographs of an ammonite and a crab shown in Figure 4. For the remaining photographs in this book, we use the second technique, which is outlined in detail in Chapter Seven.

The size of each photograph as illustrated bears little relationship to the actual size of the fossil. On each figure, the magnification is indicated by the length of the open bar in the lower left of each figure, which corresponds to 1 cm.

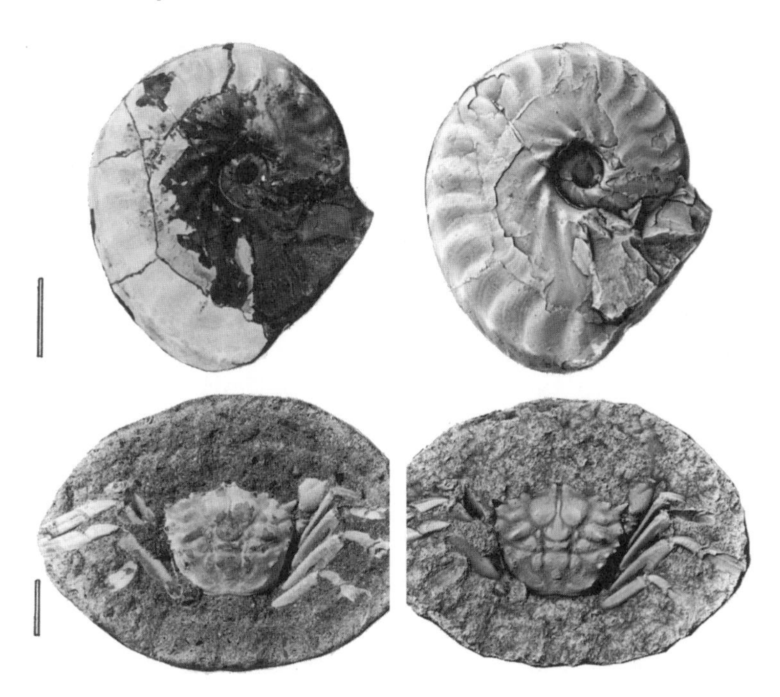

Figure 4 The ammonite *Pseudoschloenbachia* (VIPM 004) and the crab *Longusorbis* (VIPM 005), photographed as they appeared when collected (left) and again after they have been blackened and coated to emphasize surface sculpture (right).

CHAPTER TWO

Geologic Time: Measuring Eternity

"The mind seemed to grow giddy by looking so far into the abyss of time." So wrote John Playfair, Professor of Mathematics and Natural Philosophy at the University of Edinburgh in the late 1700s, when the enormity of geologic time was demonstrated to him by Dr. James Hutton at a rock outcrop on the Scottish coast. Hutton suggested that the processes that operate at almost imperceptible and exceedingly slow rates in a human lifetime—the decay and crumbling of rock, the minute shifts of river beds, the deposition of mud veneer in lagoons—had also operated in the geologic past at about the same rate. He asked rhetorically: "What would be required for these processes to form the large-scale geologic features seen in the rock record—the great thicknesses of sandstone, the tilting and erosion of rock layers, the elevation of mountain ranges?" The answer, according to Dr. Hutton: "Time; nothing but time."

Like the distance between stars, geologic time is so vast that its duration is effectively beyond human understanding. Through our grandparents, we can personally reach back nearly a hundred years. Even a thousand years can be grasped as "merely" thirty or forty

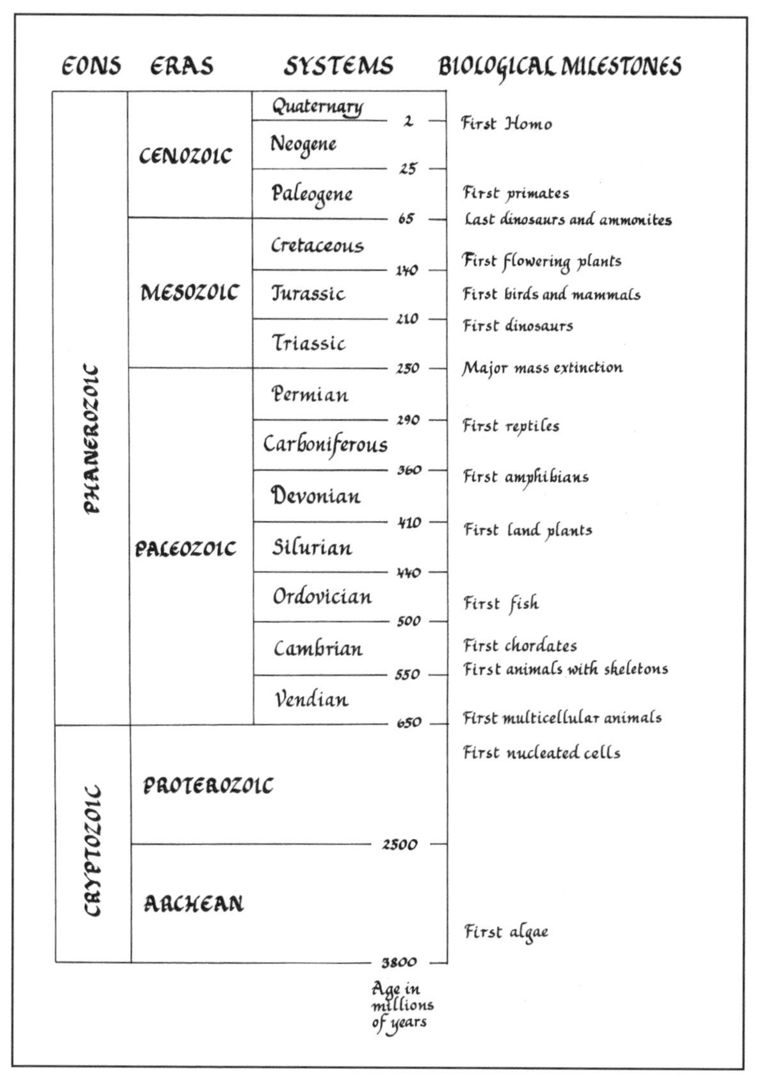

Figure 5 The geologic time scale of eons, eras, and systems showing important first and last appearances of animal and plant groups.

human generations. But can anyone truly comprehend a duration of 10 million years or a distance of 30 million km?

The oldest rocks on Earth were formed some 3,800,000,000 years ago and the star nearest to us, Proxima Centauri, is located 37,000,000,000,000 km from our sun. These are mind-numbing numbers. Geologists may substitute 3.8 Ga (giga- or billion years) for all those zeroes when giving the age of the oldest rocks, and astronomers use 4 light years for the distance of Proxima Centauri from our sun. But these numbers do not really enhance comprehension.

Immense time, however, like immense distances, can be subdivided, measured, and calibrated, if not fully understood. Geologists and paleontologists use the record of fossils in rock to partition geologic time into progressively smaller divisions—eons can be divided into eras, eras into systems, and systems into stages. Geochronologists use radioactive decay of some isotopes to measure the duration in years of these divisions (Figures 5 and 6).

The entire sweep of geologic time is subdivided into two eons—the Cryptozoic and the Phanerozoic—on the basis of the fossil record of five organic kingdoms:

MONERA—minute, simple unicellular organisms without nuclei, such as bacteria and blue-green algae (procaryotes).

PROTOCTISTA—larger, complex unicellular organisms with nuclei, such as ameboids and protists (eucaryotes).

METAZOA—true multicellular animals.

METAPHYTA—true multicellular plants.

FUNGI—multicellular moulds and mushrooms.

Rocks deposited during the first eon, called Cryptozoic (*kryptos* + *zoic*, Greek for "hidden life"), contain a sparse fossil record consisting mainly of minute and unicellular organisms of the kingdom Monera. A few members of the kingdom Protoctista appeared late in this eon. The base of the second eon, called Phanerozoic (*phaneros* + *zoic*, Greek for "evident life"), is marked by the incoming of large, multicellular organisms of the kingdoms Metazoa and Metaphyta. The kingdom Fungi has a spotty fossil record through the Phanerozoic.

THE CRYPTOZOIC EON

The Cryptozoic Eon encompasses the first 3,000 million (3 billion) years of Earth's history. It is divided into the Archean and Proterozoic eras (Figure 5). Archean rocks are exposed in the central parts of most continents. In North America, they occur in the centre of the Canadian Shield around Hudson's Bay; none occurs in British Columbia. During this era the Earth's crust was beginning to grow, but large stable continents had yet to form. The primitive atmosphere was composed mainly of gases, such as hydrogen, methane, ammonia, water vapour, and other carbon-hydrogen and nitrogen gases; significantly, it lacked oxygen. The first fossils are monerans that date from 3,600 million years ago.

A major turning point in the history of the Earth occurred some 2,500 million years ago at the beginning of the Proterozoic Era, when large, stable continents were being formed for the first time. The great abundance of algal mats in Proterozoic rocks on all continents indicate that primitive life must have proliferated in the shallow seas during this era. Biogenic oxygen produced by photosynthetic blue-green algae began to accumulate in the atmosphere, some of it contributing to an ozone screen. Sparse fossil evidence from late in the Proterozoic, a time when oxygen in the atmosphere approached 2 to 3 percent of the present levels, indicates that a new kingdom had appeared. The Protoctista are still unicellular but, unlike the procaryotic Monera, these amoeba-like organisms are eucaryotic (have a well-defined nucleus encased in a nuclear membrane), like the multicellular Metazoa and Metaphyta that were soon to develop. Proterozoic rocks are widely exposed in eastern British Columbia.

By the late Proterozoic, the continents had assembled into a single supercontinent named Rodinia. The name comes from the Russian word *rodit* meaning "to beget." Just before the end of the Proterozoic, Rodinia began to rift apart into a number of smaller continents separated by deep ocean basins. In effect, Rodinia begat all subsequent continents. One of these continents was Laurentia—geological North America.

THE PHANEROZOIC EON

An Earth of modern aspect—that is, one teeming with both microscopic oxygen-producing blue-green algae, bacteria, and ameboids, and with large animals and plants—began a scant 650 million years ago at the bottom of the Phanerozoic Eon (Figure 5). First to appear were imprints of large soft-bodied multicellular organisms in the Vendian. The quality of the fossil record improved significantly about 100 million years later with the appearance of animals having mineralized shells and skeletons in the Cambrian. Many of the animal phyla now living—Porifera, Annelida, Arthropoda, Echinodermata, Brachiopoda, Mollusca, and others—along with several extinct phyla, appeared quickly in a burst of diversification lasting little more than 10 million years in the Early Cambrian. The Burgess Shale (Middle Cambrian) in Yoho National Park in eastern British Columbia preserves evidence of this diversification: thousands of specimens of exquisitely preserved fossils belonging to a bewildering array of animal groups. With considerable justification, the Burgess Shales have been called the most important fossil site in the world, reflected in the fact that the area is now a UNESCO World Heritage Site.

The pattern of changing life through the Phanerozoic Eon gives descriptive names to three component eras (Figure 5). The seas of the Paleozoic Era ("era of ancient life") were crowded with trilobites, brachiopods, crinoids, tetracorals, and jawless fishes. By mid-era, plants began to invade the land, thus preparing the way for spiders and insects and for tetrapods, such as amphibians and reptiles, which were soon to follow. The Paleozoic Era consists of seven systems: Vendian, Cambrian, Ordovician, Silurian, Devonian, Carboniferous, and Permian. It ended with the greatest mass extinction of all time. Half the families and about 95 percent of the species of the latest Permian marine biota were terminated.

During the Mesozoic Era ("era of middle life") the seas were populated by ammonites, bivalves, hexacorals, crabs, bony fishes, and swimming reptiles. On land, dinosaurs, insects, cycads, and conifers were plentiful, and mammals, birds, and flowering plants made their first appearances. The Mesozoic Era comprises three systems—Triassic, Jurassic, and Cretaceous. Another mass extinction at the end of this era, less severe than the previous one, finished off the dinosaurs, marine reptiles, and ammonites, among other groups.

Fossil evidence of the Cenozoic Era ("era of recent life") is extensive and diverse on land, including mammals, birds, insects, and flowering plants. In the seas, bivalves, snails, crustaceans, and bony fishes continued to proliferate up to the present time. The Paleogene and Neogene systems, and the brief Quaternary, in which we are now living, make up the Cenozoic Era.

The diversification and extinction of animals and plants during the Phanerozoic Eon were strongly influenced by a dynamic global geography—a slow-motion dance of continents moving apart, rotating, and coming together as the oceans widened and closed. Late in the Proterozoic, the supercontinent Rodinia had fragmented into a number of continents. The great Cambrian expansion of animal groups occurred during this break-up. In the early Paleozoic, the continental fragments were still widely separated but, by the late Paleozoic, had converged to form the second supercontinent, Pangaea. The final assembly of this supercontinent coincided with the enormous mass extinction at the end of the Permian. Pangaea remained intact for about 70 million years, but in the middle Mesozoic it, like its predecessor, began to rift apart. The break-up of Pangaea into smaller continents strongly influenced the proliferation of animals and plants in the late Mesozoic and Cenozoic eras. Most of the continents are still separating at the present time as the intervening oceans widen at the expense of a diminishing Pacific Ocean.

CHAPTER THREE

Dating Rocks: With Fossils and by Isotopes

The visual appearance of a rock doesn't help to determine its age. Old sandstones, shales, and granites look just about the same as young sandstones, shales, and granites. Fortunately, rocks provide two different methods of age determination. The first, biostratigraphy, the study of fossils and their succession through bedded strata, contributes a relative age—that is, B is older than A and younger than C—but without indicating by how much. The second, geochronology, the measurement of naturally occurring radioactive isotopes in minerals, contributes a numerical age in years.

RELATIVE AND NUMERICAL AGES

The relative time scale of fossil-bearing rock has been in place for about 150 years. It was established empirically for the bedded fossiliferous rocks of western Europe where, for example, strata bearing Late Jurassic fossils are everywhere succeeded by strata with Early Cretaceous fossils. With the knowledge of this natural succession, a competent early Victorian naturalist (the word paleontologist had not yet been coined) would have been able to assign a European fossil collection to one of about a dozen geologic systems—Cambrian to Carboniferous to Neogene. But not only that. If the fossils were of the right kind and sufficiently well preserved, this naturalist could also assign a collection to one of perhaps a dozen stages within each system—in all, to one of about a hundred well-characterized age intervals starting in the Lower Cambrian.

The numerical time scale in years is a more recent development made possible by the discovery of natural radioactivity about a hundred years ago. This technique is based on the radioactive decay of one isotope of an element to form a different isotope; the decay occurs at a constant and known rate. The rate is expressed as a half-life in years—the time it takes for one-half of the parent isotope to decay to become the daughter isotope. The age of the mineral (and thus the rock) can be determined by the proportion of parent to daughter isotopes. The more common isotope pairs used for the numerical age determination of rocks (and their half-lives) are Uranium-238 to Lead-206 (4,468 million years); Thorium-232 to Lead-208 (14,010 million years); and Potassium-40 to Argon-40 (1,250 million years). Since 1950, geochronologists have been using this technique regularly to date Cryptozoic and Phanerozoic rocks.

AN EXAMPLE: THE AGE OF THE PUNTLEDGE ELASMOSAUR

The elasmosaur discovered in rocks on the Puntledge River in 1988 can be dated according to the two time scales described above.

The rocks along the Puntledge and Browns rivers west of Courtenay form the lower part of a thick succession of shales, sandstones, and conglomerates exposed in the Comox Valley and across Denman and Hornby islands. These strata of the Nanaimo Group are divided into seven named formations, with the Comox Formation at

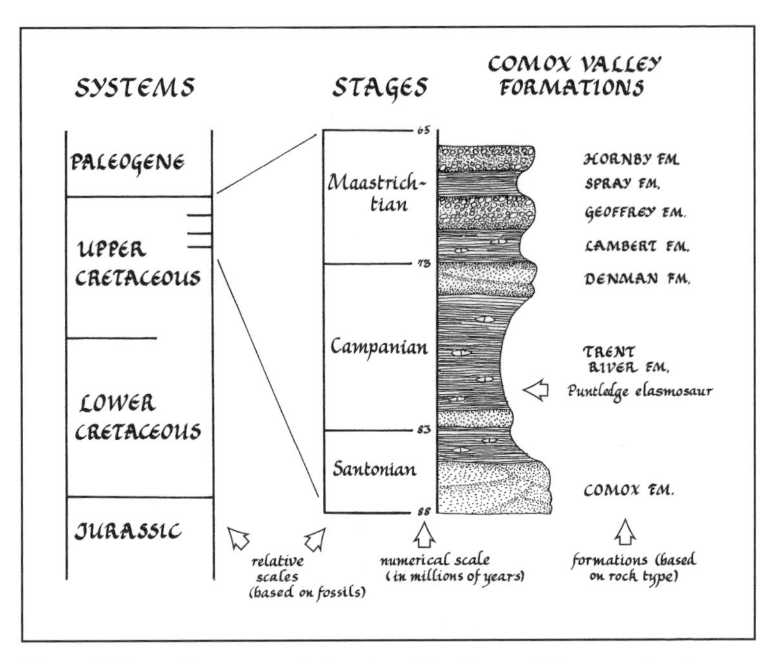

Figure 6 Upper Cretaceous stratigraphy of the Comox Valley area showing the sequence of formations. The age of the Puntledge elasmosaur can be determined according to the relative geologic time scale as Campanian and Late Cretaceous, and the fossil can be dated numerically as about 80 million years old.

the bottom and the Hornby Formation at the top (Figure 6). The sandstone and conglomerate units are virtually unfossiliferous, but fossils are common in the shale units and contain shells, bones, and teeth belonging to familiar animal groups such as clams, snails, crabs, lobsters, and sharks, and to extinct groups such as ammonites, elasmosaurs, and mosasaurs. Fossil plants are also quite common.

The rich assortment of coiled ammonites is particularly useful in indicating a relative age for these rock strata. The dozen or so genera found in the Trent River Formation exposed along the Puntledge River indicate that these soft shales are of Late Cretaceous age. In California, Alaska, Japan, the Russian Arctic, India, Australia, Madagascar, and western Europe, these ammonites are found only in the highest three stages of the Upper Cretaceous—the Santonian, Campanian, and Maastrichtian. The stages are named for localities in

France and Holland where these rocks and fossils are particularly well expressed; that is, where the strata are clearly exposed and where they contain numerous fossils.

The Puntledge elasmosaur is enclosed in Upper Cretaceous shale which, on the basis of ammonites, appears to be of late Santonian or early Campanian age according to the relative geologic time scale. The names of these stages belong in the normal vocabulary of stratigraphers and paleontologists working on Mesozoic rocks and fossils, but they mean little to other people—although the name Campanian is certainly familiar to wine aficionados as the home province of true champagne.

What about the age of the Puntledge elasmosaur in years? Radiocarbon (Carbon 14) dating is not of much use. It can only be used on organic material younger than 50,000 years. To use this technique for dating Vancouver Island rocks and fossils would be about as appropriate as using an egg-timer to record the phases of the moon.

Instead, volcanic ash provides an accurate chronometry that extends for hundreds of millions of years. And because the ash is the same age as the sediment (now rock) and the animals and plants (now fossils), it provides direct dating.

At hundreds of localities around the world (including the Puntledge River), fossil-bearing Upper Cretaceous shales, sandstones, and limestones are interleaved with thin layers of bentonite—altered volcanic ash. These ashes include minute flakes of mica minerals that can be dated numerically with the Potassium-Argon method. By systematically studying the proportion of Potassium-40 isotopes to Argon-40 isotopes in mica from literally thousands of layers of volcanic ashes, geochronologists can obtain accurate numerical ages in years, which can then be tied into the relative age scale of stages and systems.

The end of the Cretaceous was marked by a major mass extinction that eliminated every child's favourite group of fossils, the dinosaurs. It was also the end of all ammonites, all plesiosaurs (including elasmosaurs), and all mosasaurs, along with a host of other groups. This extinction has been the focus of intense paleontologic and geologic attention, particularly after the publication of an important paper in 1980 that suggested an extra-terrestrial cause—the impact of an asteroid having a diameter of 10 km and the catastrophic consequences of that asteroid's collision with Earth. The end of the Cretaceous has now been dated precisely as 65 million years ago. The stages of the Upper Cretaceous have also been dated numerically. The base of the Maastrichtian is dated as 73 million years ago and the base of the Santonian as 88 million years (Figure 6).

The fossils collected from the lower Trent River Formation exposed along the Puntledge River demonstrate that these shales are Late Cretaceous (late Santonian to early Campanian) in age according to the relative geologic time scale. The Santonian-Campanian boundary can be dated as 83 million years ago according to the numerical scale. This means that the Puntledge elasmosaur must be approximately 80 million years old.

CHAPTER FOUR

The Cordillera: New Ideas About Old Mountains

Cordillera, a Spanish word meaning "a chain or range of mountains," is the general term applied to the entire sinuous western spine of the Americas—the series of mountain ranges and chains extending from Tierra del Fuego all the way to Alaska. The Canadian portion of the Cordillera consists of the wide belt of parallel mountain ranges between the foothills of Alberta and the BC coast. Vancouver Island forms the westernmost part of the Cordillera.

MOUNTAINS AND TERRANES

Geologists are fond of mountains because mountains are about the only places on Earth where great expanses of rock can be examined and sampled. But mountains are also a source of baffling questions.

What is the nature of the tremendous forces that pushed rocks up thousands of metres above sea level? Why are mountains generally lined up in rows parallel to the edges of continents? And, aside from the obvious fact that they stick out of the ground, is there anything different about mountains or the rock of which they are made? These questions have remained fundamental problems in geology for the last 150 years.

In the 1950s and 1960s, as the deep ocean basins were being mapped for the first time, oceanographers and geologists were astonished to discover extensive submarine features, such as the mid-oceanic ridges and the deep trenches. Using the bathymetry of the deep oceans and geologic information from the continents, coupled with the pattern of earthquake and volcanic activity, geologists formulated new models of an Earth with a mobile crust. The linked models of sea-floor spreading and plate tectonics breathed new life into older speculations about continental drift.

The insights gained by these models now began to clarify the origin of mountain chains. According to these models, Earth's surface is made up of rigid plates of oceanic crust and continental crust that ride on a semi-fluid mantle and jostle each other along their margins. At some margins, such as along the mid-Atlantic Ridge, new oceanic crust is being created from the upwelling of molten rock. At other margins, for example along the San Andreas Fault in California, the plates slide past each other—sometimes with devastating consequences. At still other margins, the dense crust of an oceanic plate is shoved under thicker and lighter crust at the margin of a continent (subducted). In the latter case, frictional heat is created as rock and sediment are subducted down to deep levels where they melt and then begin to rise buoyantly through the crust as large masses of molten magma. As these masses approach the surface, the overlying rocks are uplifted, tilted, pushed aside, and faulted into linear chains of mountains parallel to the margin of the continent.

These models seem to be valid because they explain the earthquakes, the volcanic activity, and the uplift presently observed along the "ring of fire" that encircles the Pacific Ocean. The models, however, cannot by themselves explain the formation of a broad and composite mountain belt, such as the Cordillera of western North America. In

British Columbia, mountain range follows mountain range all the way from the foothills of Alberta to the west coast of Vancouver Island, an east-west distance of nearly 800 km. Each range differs from its neighbour—one is made of granite, the next of volcanics and shales, or of severely metamorphosed rocks altered by extreme heat and pressure, or of unaltered limestone, and then perhaps granite again followed by a range of volcanics with deep-water shales. Moreover, the rocks that are now juxtaposed are so different that it is inconceivable that they originally formed so close to one another. Particularly difficult to explain are large belts of shallow-water Permian limestones in central British Columbia containing the remains of organisms identical to those then living only in the tropics. How did these limestones and tropical fossils end up in the middle part of a broad composite mountain belt that apparently has remained in the northern hemisphere since the Permian? Could these animals have migrated far beyond their tropical home along narrow warm seaways within the Cordillera? This explanation did not sit well with earth scientists who had looked in vain for geological evidence of such seaways.

Then, in the early 1970s, two west coast geologists came up with a novel explanation of these disjunct limestones and fossils and, even more interesting, with a brand-new way to explain the formation of the Cordillera. Jim Monger of Vancouver and Charlie Ross of Bellingham suggested that the limestones in the Cache Creek area of central BC contain tropical fossils for the simple reason that when these rocks formed during the Permian, they were not part of Laurentia but were situated in the tropics—on and around volcanic islands then located far out in the ancient Pacific Ocean and close to the equator. Tim Tozer, then of Ottawa, made a similar suggestion for Triassic rocks and fossils of the southern Yukon Territory and Vancouver Island. These Permian and Triassic rocks and the remains of tropical organisms (by now fossils) were then separately transported as passive passengers on northward-moving sea floor, later to collide with Laurentia.

The suggestions from Monger, Ross, and Tozer—that large regions of mountainous British Columbia were formed elsewhere and later plastered onto North America as huge prefabricated rafts of rock (which they called terranes)—revolutionized the study of Cordilleran geology. The Cordillera is now seen to be a collage of terranes, each

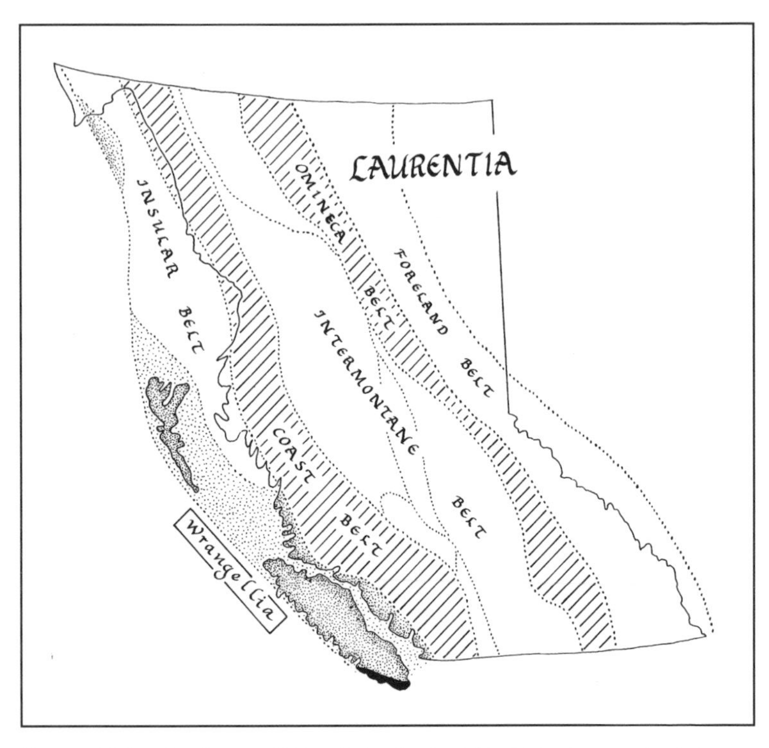

Figure 7 The Cordillera in British Columbia is divided into linear belts that express fundamental geologic divisions. The Insular and Intermontane belts consist of collages of terranes, each of which had a separate geologic history distant from Laurentia—ancestral North America. Vancouver Island, Haida Gwaii (the Queen Charlotte Islands), and parts of southeast Alaska comprise the terrane Wrangellia (stippled). The small Pacific Rim and Crescent terranes on southern Vancouver Island are shown in black. The Omineca and Coast belts (shaded) are made up largely of igneous and metamorphic rock. The Rocky Mountains (Foreland Belt) define the western margin of Laurentia. All of the belts west of the Omineca Belt were added to Laurentia during the Mesozoic.

with its own geological history. In British Columbia, some twenty major terranes have been identified to date—ranging in size from a few tens of kilometres across to as large as 1,000 km. A few of the larger terranes are outlined by dotted lines in Figure 7. The plate tectonic process by which terranes were accreted to Laurentia is now known as terrane tectonics.

MAKING BRITISH COLUMBIA

Only the eastern third of British Columbia is part of geological North America—the ancient continent Laurentia. The rest of the province is a collage—a crazy quilt of exotic terranes each made elsewhere and each with its own individual geologic history that must be unravelled separately. Most of these terranes were added to Laurentia during two periods of collision in the Mesozoic.

In the latest Proterozoic, about 800 million years ago, the first supercontinent, Rodinia, began to break up. New ocean basins were formed by sea-floor spreading as new lava welled up when the rifted continents moved away from each other. Laurentia was one of these new continents, but it would be difficult to recognize it as ancestral North America; it was considerably smaller and the ancient coastlines did not correspond to the modern ones. The eastern margin of Laurentia bisected Newfoundland and New Brunswick, and the western margin lay in what is now eastern British Columbia (close to a line joining Nelson, Prince George, and Dease Lake). The continent that rifted away from western Laurentia in the late Proterozoic is not known with certainty, but it probably was either Australia or Antarctica, or both.

For the next 500 million years, Laurentia accumulated great thicknesses of sedimentary rocks on its western margin, which remained relatively quiet with little mountain-building or volcanic activity. Not so on the other side of Laurentia, where oceans began to close in the late Paleozoic as continents collided to assemble the second supercontinent, Pangaea. Out in the ancient Pacific Ocean, volcanic arcs poured out huge volumes of lava to form numerous volcanic islands, seamounts, and plateaus. Some of these eventually drifted westward to form parts of eastern Asia. Others drifted north and east to collide later with Laurentia, forming the central and western accreted terranes of the Cordillera in British Columbia.

During the Permian, about 270 million years ago, the Cache Creek, Quesnellia, and Stikinia terranes, named for their eventual destination in central British Columbia, were clustered far out in the ancient Pacific Ocean near the equator. The Alexander Terrane and Wrangellia, which eventually formed the outboard terranes of BC and Alaska, lay deep in the southern hemisphere. During the Permian and Triassic, all of these terranes were slowly moving north. By the Early Jurassic, some 70 million years later, Quesnellia, Stikinia, and the Cache Creek Terrane amalgamated into the Intermontane Belt to form a small continent lying off Laurentia which, at the time, was part of the supercontinent Pangaea.

In the Middle Jurassic, Pangaea started to rift apart. The Atlantic Ocean began to open and, as a direct consequence, Laurentia began moving west towards the Intermontane Belt at a rate of a few centimetres a year. All intervening oceanic crust was subducted and the larger continent eventually collided with the smaller. The crust along the zone of impact thickened, compressed, heated up, and partially melted. This belt became the igneous and metamorphic rocks of the Omineca Belt now located in east-central BC. As this slow-motion collision progressed, the thick wedge of sedimentary rocks lying on the margin of Laurentia was uplifted and faulted into the Foreland Belt—the Rocky Mountains and its foothills (Figure 7).

Carrying the now-accreted Intermontane Belt on its leading edge, Laurentia continued its westward course as the Atlantic continued to widen. Some time before the Early Cretaceous, Wrangellia and the Alexander Terrane had amalgamated into the Insular Belt to form a small continent located west of Laurentia. Because of the continued westward trajectory of the large continent, a collision with the Insular Belt was inevitable. When it occurred in the Early or Middle Cretaceous, the melted crust of the collision zone formed the largest granite mass in the world—the Coast Belt of British Columbia.

By the Middle Cretaceous, about 100 million years ago, the accretion of terranes to form the Cordillera in British Columbia was essentially complete. A few small terranes were later stuffed under southern Vancouver Island, but the major events during the Cenozoic were lateral movements. The outer part of the western terrane of the Cordillera sheared off by faults and shifted northward. In this way, Wrangellia, which originally formed a single microcontinent, was separated into two pieces now located thousands of kilometres apart.

THE LIFE AND TIMES OF WRANGELLIA

It is not commonly done, but histories could be written about continents, just like they have been written about nations. Wrangellia was a discrete terrane the size of a small continent that now encompasses Vancouver Island, Haida Gwaii (the Queen Charlotte Islands), and areas in southeast Alaska. The rocks and fossils of Wrangellia tell the story of its fiery birth, its journey of thousands of kilometres, its prolonged collision with a large continent, and its final break-up into two separate pieces.

The story begins in the Devonian some 400 million years ago, far out in the ancient Pacific Ocean, when a series of explosive eruptions above a volcanic arc spewed out huge volumes of ash and lava. By the Carboniferous, when volcanism ceased, the volcanics and derived sediments had accumulated to form a coherent cluster of islands around which lime-secreting organisms began to flourish. In the Early Permian, 270 million years ago, Wrangellia consisted of a limestone-mantled volcanic plateau located in the southern part of the ancient Pacific Ocean, far to the west of the supercontinent Pangaea (Figure 8).

This volcanic plateau rifted apart some 40 million years later in the Triassic. Lava welled up through cracks and poured out across the Permian limestones to form a very thick unit of volcanics. This period of volcanism stopped late in the Triassic and once again Wrangellia began to accumulate limestone deposits, including extensive coral reefs.

Wrangellia had begun its long journey northward sometime in the Carboniferous or Permian. But this terrane did not move independently. It was part of an oceanic plate that moved, conveyor-belt fashion, in response to new crust being formed at an oceanic ridge on one margin of the plate. The paleomagnetic signature frozen into the volcanic rock as the lavas cooled indicates that by the Triassic, Wrangellia was approaching the equator from the south.

In the Early Jurassic, huge magma chambers beneath Wrangellia began to produce different types of igneous rock. First, thick accumulations of lavas and ash exploded from below and, somewhat later,

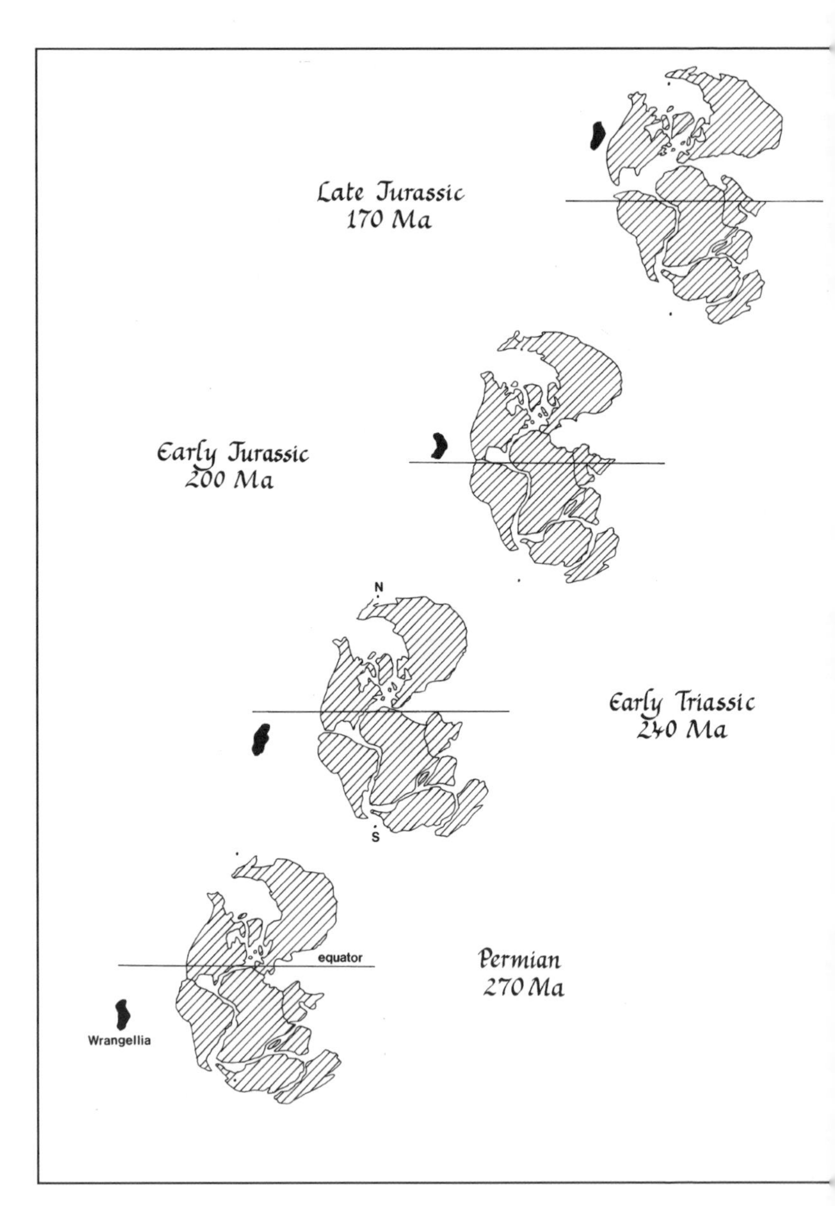

Figure 8 Starting from a position below the paleoequator in the ancient Pacific Ocean in the Carboniferous, the terrane Wrangellia drifted northward a distance of about 10,000 km and eventually collided with western Laurentia in the Cretaceous. Wrangellia is the only terrane shown on these sketch maps. The

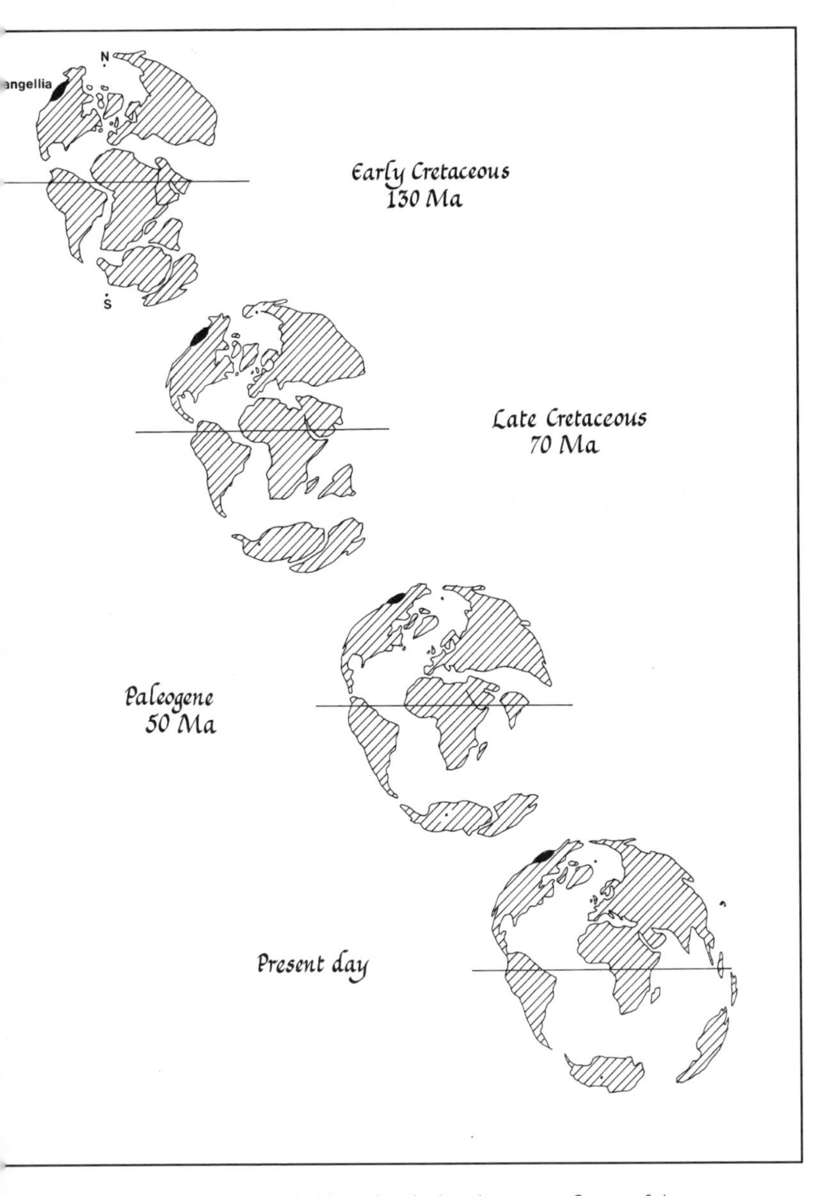

ancient Pacific was crowded by other isolated terranes. Some of these terranes eventually collided with Laurentia to form parts of the Cordillera, others formed parts of eastern Asia. The maps also depict the break-up of the supercontinent Pangaea (Ma = million years).

granitic rocks from the same magma source intruded through all the older rocks. By the Jurassic, Wrangellia had moved across the equator and the Tropic of Cancer, that is, well beyond the tropical belt where carbonate sediments prevail. As a consequence, the rocks laid down during the Cretaceous were not limestones, but shales and sandstones.

Finally, in the Early or Middle Cretaceous, after a journey of some 10,000 km that had begun about 140 million years previously, Wrangellia collided with Laurentia. This means that Wrangellia had travelled at an average speed of 7 cm a year since the Permian—about half the rate that human hair grows. This collision had far-reaching effects on the distribution of younger rocks. Uplift along the Coast Belt and in the interior of Wrangellia created a new depositional basin into which great thicknesses of shales, sandstones, and conglomerates rapidly accumulated during the Late Cretaceous.

The Upper Cretaceous rocks were later folded and faulted as two small exotic terranes were rammed under southern Vancouver Island in the Paleogene, some 40 million years ago.

Now that Wrangellia had been welded to Laurentia, the only remaining sedimentary rocks to accumulate were a narrow rim of shales and sandstones deposited along the western margin of Vancouver Island during the Late Paleogene and Early Neogene.

During the Cenozoic, as the westward-moving North American Plate slid past the northward-moving Pacific Plate, large-scale lateral dislocation of crust resulted. Wrangellia ripped in two—the western segment, now comprising parts of Alaska, shifted 2,000 km north of the eastern part, which now forms Vancouver Island and Haida Gwaii.

VANCOUVER ISLAND FORMATIONS

Throw a dart against a geological map of Vancouver Island and, more often than not, it will hit a region coloured olive green or bright red. By convention, olive green represents volcanic rock and red stands for granitic rock. Both of these igneous rock types originated as hot molten magma. The volcanics, dark-hued and fine-grained at outcrops, cooled quickly on the surface or underwater; whereas the granites,

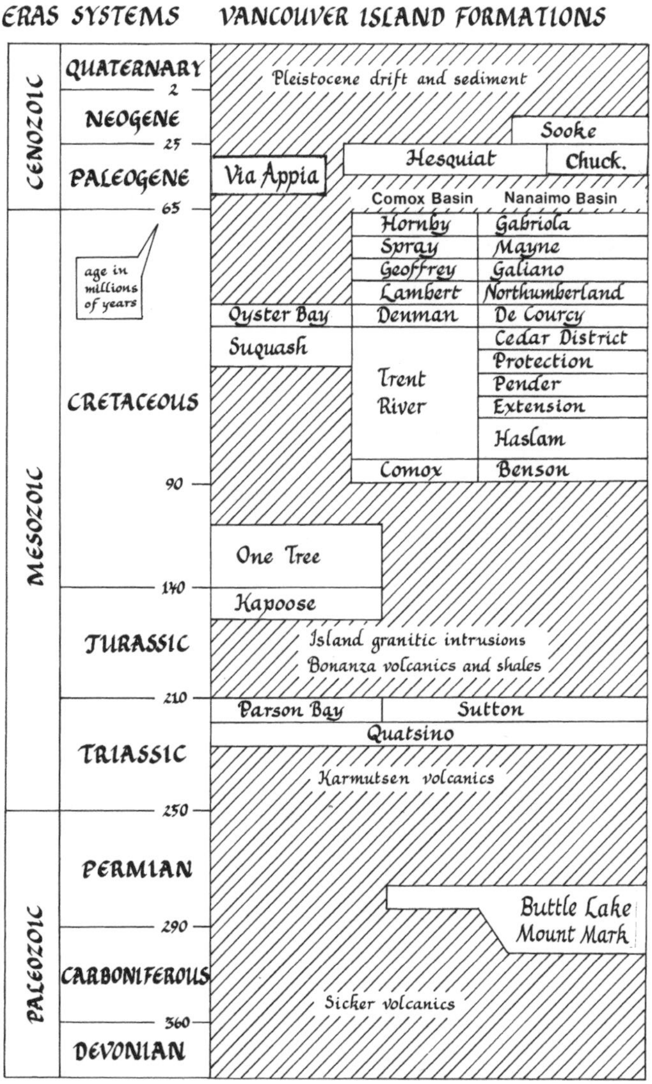

Figure 9 The formations of Vancouver Island arranged against a Devonian to Quaternary time scale. The left side of the column shows the formations on the northern half of the island; the right side shows those on the southern half. The oblique shading indicates the absence of a sedimentary rock record. Note that the Sicker, Karmutsen, and Bonanza volcanics include pockets and lenses of fossiliferous sedimentary rocks.

generally grey or pink with a salt-and-pepper texture, solidified very slowly at great depths in the Earth's crust, thus having ample time to grow large interlocking crystals. Because igneous rocks are so extensive on Vancouver Island, one might think that this island is not a particularly good place to search for fossils. But that is not the case. Sedimentary rocks, such as shales, sandstones, and limestones, are well exposed over many parts of the island. These rocks, which originated as muddy or sandy sediment on the bottom or along the margin of an ancient sea, often contain fossils that are well preserved, abundant, and sometimes startlingly beautiful.

Individual rock units occur in distinct belts, or formations, that can be mapped across Vancouver Island, sometimes for hundreds of kilometres. These formations bear familiar names because, according to rules of nomenclature, they must be named after geographic features—a lake, town, mountain, or peninsula. The formation, however, generally extends well beyond the geographic feature for which it is named. For example, shales of the Trent River Formation are exposed not only on the Trent River, but also on the Puntledge River, on Denman Island, at Northwest Bay, and at many other localities on Vancouver Island. A number of formations may be combined into a group. For example, all of the Upper Cretaceous formations on eastern Vancouver Island and the Gulf Islands are assigned to the Nanaimo Group.

Vancouver Island formations and their fossils preserve a discontinuous historical record from the Carboniferous to the Quaternary—a time interval of over 300 million years (Figure 9).

CHAPTER FIVE

Fossil Groups Represented on Vancouver Island

A great variety of fossils can be found in the sedimentary rocks exposed on Vancouver Island and surrounding smaller islands. Some of the Cretaceous and Cenozoic marine fossils—for example, many of the bivalves, gastropods, and crabs—seem so familiar that most people wouldn't give them a second glance if they found such shells washed up on present-day beaches around the Strait of Georgia. And many of the Cretaceous fossil leaves could almost have drifted down from the trees now growing in this temperate rain forest, instead of from the ancient forests covering the same ground some 70 or 80 million years ago. But other Vancouver Island fossils belong to groups that have no modern counterparts or that are extinct. Much of what we know about the biology, ecology, and evolution of groups such as ammonites, brachiopods, trilobites, tetracorals, crinoids, marine reptiles, dinosaurs, and cycads has been inferred from evidence in the fossil record.

Detailed information on each of the fossil groups preserved in the rocks of Vancouver Island can be found in recent textbooks on paleontology, such as Tidwell (1975); Murray (1985); Clarkson (1986); Boardman, Cheetham, and Rowell (1987); and Carroll (1988). Included in this book are only brief comments on the morphology and biologic meaning of the commonly occurring fossil groups.

CORALS

The phylum Cnidaria is almost entirely a marine group. It includes sea anemones, sea pens, and jellyfish, all of which lack skeletons and, therefore, have a poor fossil record. It also includes corals, which have an excellent fossil record because they secrete solid skeletons of calcium carbonate.

Two kinds of fossil corals are found on Vancouver Island. Tetracorals, also called rugose corals, possess interior vertical plates—septae—arranged in quadrants and giving a crude bilateral symmetry to the cup-like corallite. These corals, which were essential components of reefs for about 200 million years, became extinct at the end of the Permian. After a hiatus of about 10 million years, corals reappeared in the Middle Triassic, but these were hexacorals with a morphology very different from that of the extinct group. Some paleontologists think that the hexacorals must have been derived from a non-calcified sea anemone and not from the superficially similar tetracorals. The septae of hexacorals, also called scleractinian corals, impart a six-fold symmetry to the corallite.

BRYOZOANS

The phylum Bryozoa, moss animals, consists predominantly of marine colonial animals that secrete a horny or calcareous skeleton, most commonly shaped like a stick or a lacy fan. A delicate bryozoan colony is a tenement of thousands of individuals—each individual animal being ensconced in its own minute calcite box, a fraction of a millimetre in size. Bryozoans have a long fossil record extending back to the Ordovician. In virtually all cases, a thin-section and a microscope are necessary to identify a bryozoan to genus and species. A thin-section is a wafer of rock, polished so thin that it becomes transparent (generally 30 microns) and mounted between glass slides. This means that, for practical purposes, bryozoans are identified according to the form of the colony, e.g., encrusting, massive, creeping, rod-like, lacy, or spiral.

BRACHIOPODS

To marine biologists, the Brachipoda, or lamp shells, is a phylum of no great importance in present-day oceans. But to paleontologists, this phylum constitutes one of the most significant groups of marine fossils, particularly in Paleozoic rocks. Brachiopods possess two dissimilar valves that are bilaterally symmetrical. Most live permanently attached to the bottom by a horny stalk, the pedicle, which protrudes through an opening in the ventral valve. The dorsal valve supports the lophophore, the respiration and food-gathering organ of the animal. Brachiopods are rare in Mesozoic and Cenozoic rocks of Vancouver Island and, at the present time, only a single species, *Terebratalia transversa*, is found occasionally under rocks at low tides in the Strait of Georgia. In Carboniferous and Permian rocks, however, they are abundant and conspicuous fossils. Spiriferids, the most common brachiopod here, bears corrugated convex valves and a wide hinge line. Their name is derived from the internal spirals of calcium carbonate that support the lophophore. Productids, the other common group of late Paleozoic brachiopods, have flat dorsal valves and convex ventral valves. These brachiopods lack a functional pedicle and instead anchor themselves to the soft bottom with long spines.

MOLLUSKS

One out of every ten species that has ever existed on Earth is a mollusk. The phylum Mollusca includes an astonishing array of animals of virtually every shape and size—from a plain limpet to a complex pearly nautilus, and from a miniature snail less than a millimetre in size to a giant squid nearly 20 m long. Mollusks are predominantly marine animals, but a few groups occur in fresh water; one group, the gastropods, has successfully made the transition to land. Mollusks have a fossil record extending back to the Cambrian. On Vancouver Island, bivalves, gastropods, and cephalopods are frequently the only fossils encountered in Mesozoic and Cenozoic rocks.

BIVALVES

The enormous number of dead clam shells littering almost any modern beach is eloquent testimony to the outstanding success of bivalves in modern shallow seas. The shell of a bivalve consists of a pair of calcareous valves held together by interlocking teeth and sockets, and by an elastic ligament. The valves are commonly identical and closed by one or two muscles. A muscular triangular foot is generally used for burrowing, and the enfolding mantle may be fused into one or two siphons. A gland at the base of the foot secretes hair-like threads, the byssus, which serve as a means of attachment for most bivalve larvae. The byssus may be retained in some adult bivalves, such as mussels.

The life habits of living bivalves are accurately disclosed by the morphology of the shell, and so it is with fossil bivalves. Most bivalves live either within the sediment (infaunal) or on the surface (epifaunal). One can surmise the following life habits from the shape of fossil bivalves from Vancouver Island:

Infaunal burrowing: true clams and trigoniids
Epifaunal byssally attached: mussels, pen shells, and arc shells
Epifaunal attached to seaweed: paper shells
Epifaunal cementing: oysters, spiny oysters, and jingle shells
Epifaunal free-lying: inoceramids
Swimming: scallops
Rock-boring: date mussels
Wood-boring: teredo or ship worms

SCAPHOPODS

A tusk shell consists of a simple, tapered, calcareous tube that is open at both ends. This mollusk uses a pointed foot at its anterior end for burrowing into the sediment. The posterior tip of the shell opens into the water column. The Scaphopoda is an ancient group that has changed little over the past 500 million years; Ordovician scaphopods are difficult to distinguish from recent scaphopods.

The shells of recent scaphopods, better known as dentalia, were used as currency among the native peoples of the Pacific coast. Sewn into clothes or strung as necklaces, they were symbols of both wealth and power. *Antalis pretiosum*, which now lives in sandy bottoms at moderate depth, produced the best dentalia. These shells were harvested by the Quatsino people along the west coast of Vancouver Island, from where they were traded up and down the coast and across the Rocky Mountains as far inland as the Great Lakes.

GASTROPODS

Snails (gastropods) are mollusks with an elongated muscular foot used for creeping, and a well-defined head with eyes and tentacles. Most bear a single external calcareous shell that is spirally coiled, but in some snails the shell is reduced or absent. All snails possess a radula, rasp-like rows of minute mineralized or horny teeth, used in feeding. Upper Cretaceous and Cenozoic rocks of Vancouver Island contain diverse snails, including whelks, moon snails, and volutes—members of newly evolved carnivorous groups that use the radula to drill holes through the shell of their prey.

CEPHALOPODS

Cephalopods are the climax of mollusk evolution. They owe their success to a simple evolutionary novelty: the development of gas-filled chambers that provide neutral buoyancy to the swimming animal. The gas-filled chambers permitted these mollusks to leave the sea bottom for a free-swimming, itinerant existence. The shell of a cephalopod is external in nautiloids and ammonites, and internal in cuttlefish; in most squid and octopuses, it has been lost entirely.

Cephalopods are complex invertebrate animals with a head bearing prominent eyes and a mouth surrounded by tentacles. These mobile marine predators and scavengers possess a highly developed nervous system, including a true brain, and their jaws consist of upper and lower interlocking beaks. Living cephalopods include the pearly *Nautilus*, the cuttlefish *Sepia*, the squid *Loligo*, the giant squid *Architeuthis*, and the octopus.

Even though they are widespread in today's oceans, cephalopods are predominantly a fossil group; more than ten thousand extinct species have been described, compared to fewer than one thousand living species. Mesozoic rocks of Vancouver Island contain important representatives of nautiloids, ammonites, belemnites, and sepiids.

In nautiloids and ammonites, the external shell is divided into a phragmocone—the portion of the shell containing the gas-filled chambers—and a chamber occupied by the body of the living animal. The septae are the wall-like divisions that separate the chambers. A porous tube called the siphuncle connects the chambers. The shell is most commonly planispirally coiled—either as evolute shells, in which the previous coils (or whorls) are fully exposed, or as involute shells, in which the later coils cover the first-formed coils.

Nautiloids. The pearly *Nautilus* is the sole living representative of the great group of cephaloids having chambered external shells that had appeared in the Late Cambrian. For this reason, paleontologists have studied this East Pacific denizen intensely for clues about how these largely extinct animals lived. Fossil nautiloids have straight, curved, or coiled shells whose septae join the external shell along straight or gently curved paths (called sutures). The siphuncle is located in a central position. Living *Nautilus*, like the fossil nautiloids of Vancouver Island, is involute and lacks prominent sculpture on the shell.

Ammonites. Ammonites are similar to nautiloids, but they possess sutures that are strongly deflected, fluted, or crenulated and a siphuncle that is located along the outside margin of the shell, the venter. Unlike nautiloids, which generally lack surface sculpture, ammonites frequently bear ribs, constrictions, knobs, or spines, and many compressed ammonites have a raised keel.

These fossil objects were known to the ancient Greeks, who noted their resemblance to coiled ram's horns—the sacred symbol of the god Ammon. They became known as "the horns of Ammon" and eventually ammonites.

Most ammonites are planispirally coiled, with each whorl in contact with and overlapping previous whorls. Other ammonites, called heteromorphs, depart from this compact morphology, having shells of rather aberrant shapes. In some heteromorphs, the coiling of the initial shell is snail-like, followed by an upwardly recurved living chamber. In others, the initial few coils are followed by a straight shell, or a shell with two, three, or four parallel shafts—like a paper clip. Heteromorphs occur in both the Triassic and Jurassic, but are particularly well developed in the Late Cretaceous. Some of the best examples of heteromorphs come from Vancouver Island and the Gulf Islands.

The boom and bust evolutionary history of the ammonites lasted for 300 million years between the Early Devonian and the Late Cretaceous. The group nearly suffered extinction three times—at the end of the Devonian, the end of the Permian, and the end of the Triassic—before their complete demise at the end of the Cretaceous. Only a small handful of ammonites survived each of these nearly complete extinctions to give rise to successive rapid diversifications in the Carboniferous, the Triassic, and the Jurassic-Cretaceous. The mass extinction at the end of the Cretaceous was similar to the three earlier extinction intervals, but it was complete and it was final.

Belemnites. Curious cigar-shaped fossils occur in some Jurassic and Lower Cretaceous rocks of Vancouver Island. Because these objects include a conical chambered shell with a siphuncle, they are interpreted to be the internal shell of extinct squid-like cephalopods: belemnites. Their name comes from the Greek "belemnon," meaning dart or javelin. The solid shell, which consists of radiating crystals of calcium carbonate, probably acted as a counterweight to ensure that the belemnite squid remained level while swimming.

Sepiids. The recent cuttlefish, *Sepia*, has a flat, internal, chambered calcareous shell, the cuttlebone. This shell is comparable to the phragmocone of nautiloids or ammonites and it served the same function: maintenance of neutral buoyancy. Fossil cuttlebones are quite rare, but a few specimens of two different forms have been collected from Upper Cretaceous concretions on Hornby Island.

ARTHROPODS

At the present time, the Arthropoda is the most diverse animal phylum on Earth. Because all arthropods possess exoskeletons, which may be rigid and mineralized, their fossil record is generally good. Unlike all other animals, arthropods must shed their outer skeleton in order to grow. This means that many arthropod fossils represent moults, not carcasses. The name arthropod identifies the most characteristic feature of these animals: their jointed legs.

Of the four classes of the phylum Arthropoda currently recognized (Trilobita, Chelicerata, Crustacea, and Uniramia), only the chelicerates (horseshoe crabs, spiders, mites, and scorpions) have not yet been found as fossils on Vancouver Island.

TRILOBITES

Trilobites are found only in marine rocks of Cambrian through Permian ages. A trilobite possesses a three-fold division of the body into an anterior cephalon, a middle thorax, and a posterior pygidium, but it derives its name from a longitudinal division of the body into a central axial lobe flanked by a pair of pleural lobes. Most trilobites have paired compound eyes. During moulting, the cephalon separated along lines of weakness (called facial sutures) into a cranidium and two free cheeks.

Trilobites are among the most eagerly sought-after fossils. In spite of repeated searches of Carboniferous and Permian rocks on Vancouver Island over many years (and to the considerable annoyance of the head of the Denman Institute for Research on Trilobites!), these fossils eluded capture. Finally, a few years ago, the persistent efforts of a diligent paleontologist paid off when he discovered two kinds of trilobites in Carboniferous rocks near Alberni Inlet.

CRUSTACEANS

Crustaceans are a very large group of dominantly aquatic arthropods that includes such different animals as crabs, lobsters, shrimp, barnacles, and wood lice. Because they bear five pairs of appendages on the thorax, one of which is modified into grasping pinchers, crabs, shrimp, and lobsters are assigned to the order Decapoda. Lobsters have a laterally compressed carapace and a long abdomen that ends in a tail-fan. Crabs have a flattened carapace and a reduced abdomen that is permanently bent under the thorax. Barnacles are atypical crustaceans that build an external shell of overlapping calcareous plates. Wood lice are the only truly terrestrial crustaceans.

UNIRAMIANS

The class Uniramia includes insects and centipedes. Insects are represented by more than a million living species and, despite being small and delicate, they have a rich fossil record extending back to the Devonian. On Vancouver Island, however, they are exceedingly rare fossils. A beetle wing cover, a cockroach wing, and a termite nest in marine rocks are mute records of a few of the insects that inhabited the forests surrounding the embayment of the sea during the Late Cretaceous.

ECHINODERMS

Living echinoderms include sea stars, brittle stars, sea lilies, sea urchins, sand dollars, and sea cucumbers. All echinoderms possess a solid or flexible internal skeleton of porous calcite plates covered by a spiny skin. Most display a characteristic five-fold symmetry. The phylum Echinodermata is poorly represented among the fossils of Vancouver Island. Stalked crinoids are very common in Permian limestones, but they cannot be identified to genus or species because only the stalk ossicles—shaped like poker chips—are represented, not the diagnostic intact crowns. A single minute crown of a blastoid is known from Carboniferous rocks. Two species of free-swimming crinoids occur in Cretaceous rocks. Asteroids (sea stars) and ophiuroids (brittle stars) are known from one or two occurrences each. Sea urchins (regular echinoids) are extremely rare fossils, as are heart urchins (irregular echinoids), which have a bun-shaped body with bilateral symmetry superimposed on five-fold radial symmetry.

FISHES

Fish teeth, scales, fins, and bones are common fossils in many Cretaceous formations on Vancouver Island, but complete articulated specimens have not yet been collected. The fossil fishes belong to two groups: the cartilaginous fishes (sharks, skates, and rays) are represented by abundant and diverse assemblages of isolated teeth and vertebrae, and the bony fishes (most modern fishes, such as perch, salmon, eel, and seahorse) are represented by scales, fused fins, and, occasionally, bone. Although more difficult to assign to a specific group, coprolites—fossil excrement—of both sharks and bony fishes also occur in Cretaceous rocks.

Nearly complete articulated specimens of primitive ray-finned fishes have recently been discovered in Triassic strata on the Keogh River in northern Vancouver Island.

CONODONTS

When Paleozoic and Triassic limestones are dissolved in acetic acid (vinegar), the residue often includes numerous peculiar tooth-like elements composed, like our own teeth, of the mineral apatite (calcium phosphate). These micro-fossils (less than a millimetre long), collectively called conodonts, have been intensely studied because they are extremely useful for dating and correlating rock successions. The identity of the conodont animal, however, remained unknown until the early 1980s, when a slab with a 4-cm-long worm-like fossil was discovered in a museum drawer in Edinburgh. The head of this soft-bodied fossil included a pair of eyes; the tail carried low fins. Most importantly, it included a number of conodont elements in the mouth region. This specimen demonstrates that conodonts are the fossil remains of jawless fishes, similar to living hagfish and lamprey, that became extinct at the end of the Triassic.

MARINE REPTILES

Until recently, fossil vertebrates other than fish were not known from the Vancouver Island Cretaceous. Over the past decade, however, a number of discoveries of bones and teeth of three different types of large marine reptiles point to the existence of a previously unknown vertebrate fossil region in western Canada. And, in a striking demonstration that new important paleontological players have arrived, all of these discoveries have been made by amateurs. Many of these fossils came from the lower part of the Trent River Formation in the Comox Valley. Others came from the Lambert Formation on Hornby Island and from the Trent River Formation on Englishman River. Well-preserved fossil bones and teeth of elasmosaurs, or swan lizards, have been collected, including a moderately complete skeleton 10 m long. Mosasaurs, also known as sea lizards, are represented by vertebrae, teeth, and by their bite marks on ammonites. Fossil bones of the marine turtle *Desmatochelys* have also been discovered.

On Vancouver Island, fossils of marine reptiles have only been found in Cretaceous rocks. The single exception is a large, poorly preserved ichthyosaur, or fish lizard, that was collected from Upper Triassic limestones along Holberg Inlet on the northern part of the island.

PTEROSAURS

In the Late Triassic, pterosaurs became the first vertebrate animals to take to the air in active flight. These flying reptiles were joined by birds in the Jurassic, and they were followed by bats (mammals) in the Paleogene. The wing membrane of pterosaurs was supported by a greatly extended fourth digit of the hand. Among other anatomical modifications, the development of thin-walled and hollow bones that were easily broken after death is important to paleontologists because it resulted in a general scarcity of well-preserved fossil specimens. A single hand bone from the Upper Cretaceous of Hornby Island appears to be the sole fossil record of pterosaurs west of the Rocky Mountains. Along with dinosaurs and marine reptiles, pterosaurs became extinct at the top of the Cretaceous.

DINOSAURS

The mountains of British Columbia have not been kind to dinosaurs. It is known that they lived here. The Peace River area of the province does contain rich and diverse dinosaur footprints and trackways of Early Cretaceous age (or rather, contained, because most of these trace fossils disappeared beneath the waters of Williston Reservoir rising behind the Bennett Dam). But, in striking contrast to neighbouring Alberta, which includes some of the richest dinosaur localities in the world, virtually no dinosaur bones have been found in Cretaceous rocks of British Columbia.

The discovery of a single small tooth of a theropod dinosaur in the Trent River Formation now demonstrates that dinosaurs lived along the shores of the Pacific Ocean during the Late Cretaceous. The Theropoda is a group of meat-eating dinosaurs that includes the archetypal predatory dinosaur *Tyrannosaurus*.

MAMMALS

Mammals, either placental or marsupial, are geological newcomers that have only been around since the Late Jurassic, that is, for less than 150 million years. Their fossil record in the Cenozoic rocks of Vancouver Island (and elsewhere) consists largely of teeth. One paleontological wag has commented that the evolutionary history of mammals appears to be one of teeth interbreeding with teeth to produce slightly modified descendant teeth. With the exception of a single whale skull and a partial jaw of an extinct carnivore, the only fossils found on Vancouver Island are teeth of amphibious marine mammals in Cenozoic sandstones near Sooke.

VASCULAR PLANTS

Fossil and living vascular plants (that is, those with water- and food-conducting tissues) are assigned to three great groups: spore-bearing plants, gymnosperms, and angiosperms. Plants that reproduce by spores first appeared in the Silurian; today they include true ferns, club mosses, and horsetails. Gymnosperms, those plants bearing exposed seeds and pollen, developed by Devonian times; these include conifers, the maidenhair tree, and cycads. Angiosperms, flowering plants in which the seeds are protected within an ovary, first appeared in the Early Cretaceous. The development of flowers and fruits greatly aided the distribution of seeds so that by the Late Cretaceous, angiosperms had diversified into many land and water habitats. Today, with a quarter of a million species, they are the most successful plants.

Angiosperms are, in turn, divided into monocots and dicots. In monocots, the floral parts occur in groups of three, the leaves are parallel-veined, and the vascular bundles are scattered throughout the wood. Monocots include palms, lilies, orchids, and grasses. In dicots, the floral parts occur in groups of four or five, the leaves have net-like veins, and the vascular bundles are arranged in cylinders around the pith layer. Dicots include broad-leafed trees, shrubs, herbaceous forms, and most flowers.

Terrestrial plant material and wood debris are surprisingly common in the Upper Cretaceous and Cenozoic marine formations on Vancouver Island and the Gulf Islands. In the Lambert Formation on Hornby Island, for example, concretions with palm leaf-fans and permineralized wood occur immediately next to concretions with ammonites and crabs. At other localities, coalified chunks of teredo-bored driftwood and carbonaceous twigs and leaves represent plant debris that must have been swept out to sea.

An important new Cretaceous plant site on Vancouver Island was discovered in August 1996 by an observant bulldozer operator, who spotted large fossil plants while moving blasted rock of the Protection Formation at the approach to the new ferry terminal at Duke Point, south of Nanaimo. This site, named for the nearby Cranberry Arms Inn, has yielded exceptionally well-preserved plants, including palm fronds of biblical proportions, as well as flowers of a variety of plants.

Reproductive structures are particularly important for paleobotanical research, so the recent discovery of abundant three-dimensional and permineralized seeds, fruits, and nuts in Paleogene concretions in the Via Appia Beds south of Campbell River is important for Cenozoic paleobotany of the West Coast.

Many Cretaceous fossil leaves seem remarkably similar to leaves on living plants. However, except in a few cases, such as the maidenhair tree (*Ginkgo*) and the sycamore (*Platanus*), it is unlikely that they are closely related to recent plants. In addition, because different trees can bear essentially identical leaves, it is very difficult to identify, with any confidence, plant genera and species from foliage alone.

CHAPTER SIX

The Fossils of Vancouver Island

This chapter includes most of the common fossils found in the sedimentary rocks of Vancouver Island, as well as a number of the rarer forms. The islands in the Strait of Georgia are also included, as are Quadra Island to the north, and the San Juan Islands in Washington State. These fossils are dealt with in five sections—one for each of the major geologic time divisions—arranged stratigraphically from oldest to youngest.

CARBONIFEROUS AND PERMIAN

Between Duncan and Chemainus, the Island Highway slices through a series of small hills of greenish volcanic rocks. These volcanics, the oldest rocks and the true foundations of the terrane named Wrangellia, began to accumulate early in the Devonian. By the Carboniferous, they had piled up to thicknesses of more than 5 km. They take their name from Mount Sicker near Duncan.

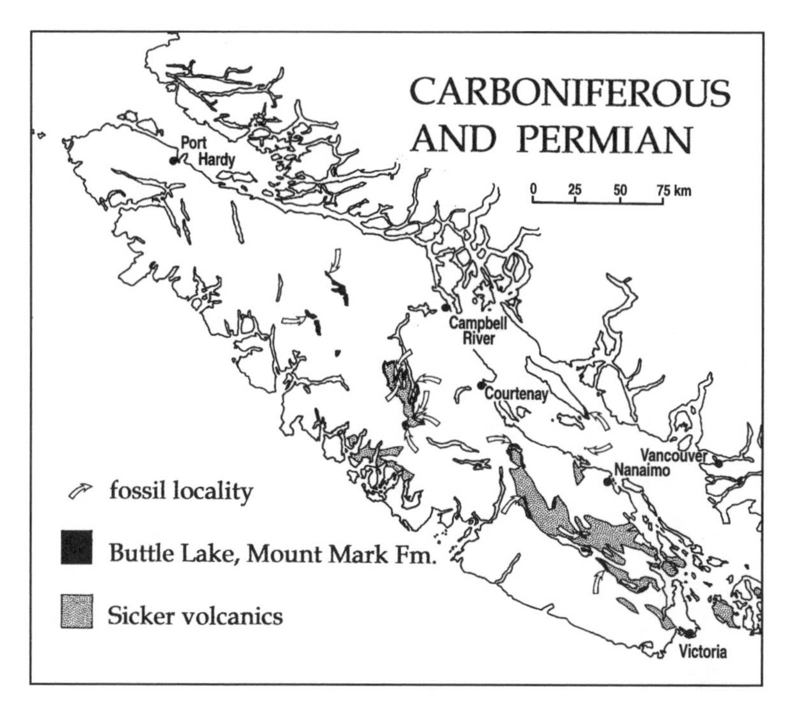

Figure 10 Distribution of Carboniferous and Permian rocks on Vancouver Island and fossil localities.

The Sicker volcanics are exposed in three main areas: in a belt extending from Horne Lake through Cowichan Lake to Saltspring Island, at Nanoose north of Nanaimo, and in the Buttle Lake area (Figure 10). Because they are the oldest rocks on Vancouver Island and the deepest buried, the Sicker rocks are also the poorest preserved. Their long, tortured geologic history can be surmised from the recrystallization and deformation evident in most outcrops. The Sicker includes some sedimentary rocks, but fossils other than microfossils and poorly preserved plant debris have not yet been collected from this thick unit.

By the Carboniferous, volcanism ceased and soon lime muds were being deposited on Wrangellia and rich associations of marine organisms began to flourish. The muddy limestones and shales of the Mount Mark Formation contain diverse brachiopods, corals, bivalves, bryozoans, blastoids, and trilobites. The calcium carbonate shells of these organisms have been dissolved so that the fossils can only be seen on weathered surfaces, where they are visible as external and internal molds. With a Late Carboniferous age, these are the oldest diverse marine fossils on Vancouver Island.

Figure 11 Life on the sea bottom during deposition of the Buttle Lake Formation exposed at Marble Meadows in Strathcona Provincial Park. Dense meadows of waving crinoids strain food particles from the sea water above fan-shaped bryozoans and clusters of spiriferid brachiopods, productid brachiopods, and rugose corals (artist Tina Beard).

Somewhat later, lime-secreting organisms began to flourish in the shallow, warm tropical seas. White and grey limestones of the Buttle Lake Formation containing common fossils of Early Permian age are widely distributed on Vancouver Island. The massive white limestone beds consist of large interlocking calcite crystals with poorly preserved brachiopods and recrystallized crinoid ossicles. By contrast, the thin-bedded grey limestones contain abundant, diverse, and well-preserved fossils dominated by large brachiopods and lacy bryozoans. Both types of limestone contain irregular nodules of fine-grained silica called chert, or flint. Silica has also replaced the shell of many of the fossils. These specimens tend to be more resistant to weathering and, therefore, protrude from the surface of the limestone. The delicate nature of many of these fossils indicates that these animals must have lived in shallow, quiet lagoons (Figure 11).

The best and most fossiliferous exposures of the Mount Mark Formation are found at Rift Creek near Mount Spencer south of Port Alberni. Good exposures of the Buttle Lake Formation are at Marble Meadows west of Buttle Lake in Strathcona Provincial Park and at Horne Lake, west of Qualicum Beach.

The Permian System, named for the Russian city of Perm, was the twilight of the Paleozoic world. For about 200 million years, starting in the Ordovician, the marine realm had been dominated by brachiopods, trilobites, nautiloids, crinoids, rugose corals, and bryozoans. Most of these groups were decimated and some became abruptly extinct some 250 million years ago at the greatest of all mass extinctions. At the end of the Permian, fully half of all animal families, and perhaps 95 percent of marine species, became extinct over a few million years. These doomed animal groups include most of the common fossils found in the Mount Mark and Buttle Lake formations—spiriferid and productid brachiopods, rugose corals, stalked crinoids, blastoids, trilobites, and bryozoans.

None of the Carboniferous or Permian faunas of Vancouver Island (with the exception of conodonts) has yet been fully described to the species level, so few of these fossils can be confidently identified.

PUBLISHED REFERENCES—Carboniferous and Permian

Rocks—Massey and Friday (1987), Muller (1980), Yole (1969).

Fossils—Fritz (1932), Yole (1963).

BLASTOIDS

Agmoblastus (Figure 12). Blastoids are stalked echinoderms that look superficially like crinoids, but have a small bud-like calyx carrying numerous thread-like arms (rarely preserved) and a pronounced pentagonal outline. All blastoids became extinct near the top of the Permian. *Agmoblastus* n. sp. is a rare fossil in the Mount Mark Formation (Late Carboniferous) at Rift Creek. The single available

Figure 12 The blastoid *Agmoblastus* n. sp. (B. Hessin Collection). Mount Mark Fm., Rift Creek.

specimen reveals a highly complex animal. The upper side of its five-sided calyx contains a central mouth surrounded by five spines, which, on closer inspection, are seen to be strongly modified: one is a curved hood covering the anal opening; each of the remaining four spines contains paired openings that are the outlets of the respiratory system.

Figure 13 Crinoid ossicles (CDM) and bryozoans (RBCM). Both from Buttle Lake Fm., Marble Meadows, Strathcona Park.

CRINOIDS

Crinoid ossicles (Figure 13). Like other echinoderms, a crinoid has an internal skeleton of porous calcite plates that is covered by a spiny skin. Large poker-chip-shaped stem ossicles are abundant in the Buttle Lake limestones, suggesting that the sea floor must have been densely covered by a veritable meadow of flexible, metre-high stalked crinoids (see Figure 11). These rooted animals wafted back and forth in the water currents using flexible arms on their calyx (head) to strain food from sea water. The plates and ossicles are difficult to assess because a well-preserved calyx is needed to identify a crinoid to genus and species. Calyces have not yet been found in the Buttle Lake limestones because they tend to fall apart when the animals die. The identity of these crinoids remains unresolved.

BRYOZOANS

Bryozoans (Figure 13). Fan-shaped bryozoans are very common in the Buttle Lake limestones. During life, these colonies faced into the prevailing current, deflecting the moving water through small, regular holes, enabling the individual bryozoans to filter out food with their lophophores. Stick-shaped bryozoans are also common.

BRACHIOPODS

Neospirifer (Figure 14). During the Silurian, spiriferid brachiopods became important members of shallow marine communities, where they flourished until their extinction at the end of the Permian. *Neospirifer* sp., the most common brachiopod in the Buttle Lake limestones, is a typical spiriferid brachiopod. It has a wide hinge line that extends for the full width of the shell, a shallow median furrow on the ventral valve, and a corresponding fold on the dorsal valve. The valves are covered by finely incised ribs that tend to occur in bundles. During life, the animal was permanently anchored to the sea bottom by a horny stalk that protruded through a triangular opening of the ventral valve.

Figure 14 The brachiopod *Neospirifer* sp. (RBCM). All from Buttle Lake Fm., Marble Meadows, Strathcona Park.

Septospirifer (Figure 15). According to Chinese mythology, spiriferid brachiopods such as this were "stone swallows" that flew about during thunderstorms. *Septospirifer* sp. from the Buttle Lake limestones at Marble Meadows has extremely wide and pointed "wings" and is evenly covered by fine radiating ribs. The low fold and sulcus have sharply defined edges.

Figure 15 The brachiopods *Septospirifer* sp. (RBCM) and *Brachythyris* sp. (RBCM). Both from Buttle Lake Fm., Marble Meadows, Strathcona Park.

Brachythyris (Figure 15). This large spiriferid brachiopod is diamond-shaped in outline, with a rather short hinge line. The ribs are few and rounded. *Brachythyris* sp. is an uncommon brachiopod in the Buttle Lake limestones at Marble Meadows.

Figure 16 The brachipod *Kochiproductus* cf. *porrectus* (RBCM). Both from Buttle Lake Fm., Marble Meadows, Strathcona Park.

Kochiproductus (Figure 16). Productid brachiopods are strikingly different from spiriferid brachiopods and they lived in an entirely different way. *Kochiproductus* cf. *porrectus* from the Buttle Lake limestones is a particularly large species, but otherwise is a typical productid. This widely distributed brachiopod was first described from Permian rocks on Greenland and named after the Danish arctic explorer, Lauge Koch. The ventral valve is strongly convex and the dorsal valve is gently concave. It lived, ventral valve down, almost entirely buried in the mud with only a crescent-shaped rim showing. Both valves bear a sculpture of closely spaced rows of nodes which, in life, were the attachment points of long hollow spines. Productid brachiopods were the only brachiopods that lived buried in the sediment (infaunally). All other brachiopods lived on top of the sediment (epifaunally). A living *Kochiproductus*, therefore, would have been covered by a dense, hair-like mane of fine spines, which served to anchor it in the mud.

Figure 17 The brachiopods *Chonetinella* sp. (top), *Krotovia* sp., and *Reticulatia* sp. (B. Hessin Collection). All from Mount Mark Fm., Rift Creek.

Reticulatia (Figure 17). As its name implies, this productid brachiopod is covered by a network of fine furrows that intersect at right angles to define regular rows of small nodes. Spine bases are sparsely distributed on both valves. *Reticulatia* sp. has a highly convex and bilobate ventral valve and a gently concave dorsal valve. It occurs in the Mount Mark Formation (Upper Carboniferous) at Rift Creek.

Krotovia (Figure 17). A fat little productid brachiopod, *Krotovia* is densely covered by spine bases and, in life, must have looked like a miniature pincushion. It occurs in the Mount Mark Formation at Rift Creek.

Chonetinella (Figure 17). A few chonetid brachiopods are found in the Mount Mark Formation at Rift Creek. Chonetids were the ancestors of productids from which they differ mainly by lacking spines over the shell. *Chonetinella* sp. is conspicuously bilobate with a wide hinge line and fine radiating ribs. Along with the spiriferid and productid brachiopods, chonetids became extinct at the end of the Permian.

Derbyia (Figure 18). This rare brachiopod has thin, nearly flat valves that are radially ribbed and fan-shaped. *Derbyia* may have lived attached to the stems of crinoids or to erect bryozoans.

Figure 18 The bivalve *Ptychopteria* sp. (RBCM), the brachiopod *Derbyia* sp. (RBCM), and the rugose coral *Caninia* sp. (RBCM). All from Buttle Lake Fm., Marble Meadows, Strathcona Park.

BIVALVES

Ptychopteria (Figure 18). The fauna of the Buttle Lake Formation is dominated by crinoids, bryozoans, and brachiopods; bivalves are extremely rare. *Ptychopteria* sp. is a small bivalve having fine, densely spaced growth lines that lived byssally attached above the bottom, possibly to erect bryozoan colonies.

TETRACORALS

Caninia (Figure 18). Only a few poorly preserved solitary tetracorals occur in the Buttle Lake limestones. The illustrated specimen is deeply eroded; the parallel ridges along the flanks are traces of the interior septae.

TRILOBITES

Paladin (Figures 19 and 20). In many features, *Paladin* is a standard trilobite that would not seem out of place in Ordovician or Devonian rocks. It is a compact trilobite with a semicircular head and tail, lacking spines or tubercles, and rather large eyes set close to a prominent smooth glabella. In the final decline of trilobites in the Carboniferous and Permian, *Paladin* had a global distribution. Complete specimens of *Paladin* sp. are not known from Vancouver Island, but a reconstruction is possible. A single collection from the Mount Mark Formation at Rift

Figure 19 The trilobite *Paladin* sp., free cheek, pygidium and close-up of the visual surface of the compound eye (B. Hessin Collection). Mount Mark Fm., Rift Creek.

Creek includes a couple of well-preserved pygidia and free cheeks, but only a poorly preserved thorax and cranidium. One free cheek is of considerable interest because it preserves the curved visual surface of a large compound eye comprising hundreds of closely packed lenses. These lenses were of calcite; trilobites being the only animals to use this mineral as an optical medium. Trilobite eyes such as this had first appeared in the Early Cambrian, 250 million years previously, so they were already ancient by the Late Carboniferous.

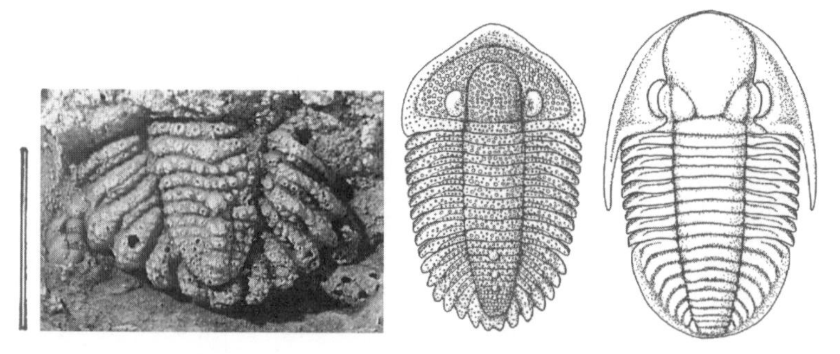

Figure 20 The trilobite *Brachymetopus* cf. *pseudometopina*, pygidium (B. Hessin Collection). Mount Mark Fm., Rift Creek. Bill Hessin's reconstructions of the two trilobites from the Mount Mark Formation.

Brachymetopus (Figure 20). A few exquisitely preserved pygidia from the Mount Mark Formation can be assigned to *Brachymetopus* cf. *pseudometopina*. The pygidium is deeply furrowed and ornamented with coarse tubercles. Larger tubercles run along the mid-line. The rear margin is a series of blunt spines. A fragment shows that the front of the cephalon comes to a blunt point. Because *Brachymetopus* has fused facial sutures, the cephalon was shed as a single unit during moulting. *Brachymetopus* is one of the longest ranging trilobite genera known—lasting 100 million years through the Carboniferous and Permian.

Figure 21 The conodont *Gondolelloides canadensis* (GSC 94281), upper and lower view; specimen is 1 mm long. Mount Mark Fm., Cameron River. The conodont stamp issued by Canada Post in 1991. *G. canadensis* is the specimen on the upper left.

CONODONTS

Gondolelloides (Figure 21). In the early 1990s, Canada Post issued an attractive series of stamps featuring Canadian fossils, including one with five different kinds of Late Paleozoic conodonts. The elongate knobby conodont shown at the top of this stamp is a widespread species, *Gondolelloides canadensis*, which occurs in the Lower Permian part of the Mount Mark Formation on Vancouver Island, and widely in western and arctic Canada. This slender, gondola-shaped conodont is 1 mm long. It has a deep slit on the lower side and is ornamented by a row of transversely flattened nodes on the upper side. Like many other microfossils, conodonts are best pictured by scanning electron micrographs.

TRIASSIC

Mount Arrowsmith above Port Alberni is the kind of mountain that children are fond of drawing—isolated and steep-sided, with grey-purple jagged peaks that retain some ice and snow even in August.

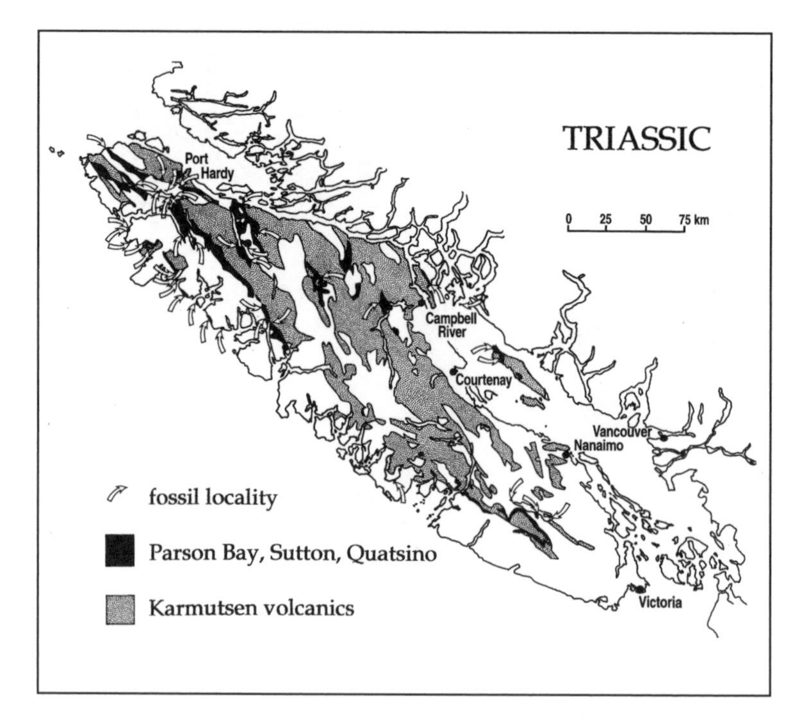

Figure 22 Distribution of Triassic rocks on Vancouver Island and fossil localities.

This mountain is made of layer upon layer of ancient lava that poured out onto the sea floor some 230 million years ago during the Triassic.

There's a lot of Triassic rock on Vancouver Island—in fact, more rock of Triassic age than of any other geologic system (Figure 22). Unfortunately (at least for the paleontologist and the fossil collector), virtually all of it is volcanic in origin. These brown and green weathering volcanic rocks are assigned to the Karmutsen Formation. Karmutsen, incidentally, was not named for a Norwegian settler; instead, it is a corruption of the Kwak'wala word *karmutzen*, meaning "falls"—the name for Nimpkish Lake on northern Vancouver Island where these rocks are well exposed. The Karmutsen comprises thousands of metres of volcanics that piled up on the sea floor over a 5-million-year period in the Middle Triassic. These volcanic rocks consist of densely packed pillows of lava, basalt flows, and ash beds. A few thin interbeds of fossil-bearing limestone and shale demonstrate that bivalves and ammonites flourished locally while this submarine lava field spewed out.

The paleomagnetic record that was frozen into these volcanics as the lava cooled indicates that Wrangellia was located near the equator during the Triassic—far out in the ancient Pacific Ocean and far away from the supercontinent Pangaea.

Late in the Triassic, volcanism finally ceased on Wrangellia, to be replaced by widespread deposition of light grey to black limestones of the Quatsino Formation. A few hundred metres of fine-grained limestone accumulated, some massive and some with interbedded black chert as beds and nodules. Much of the limestone was later crinkled into tight folds, for example, at Open Bay on Quadra Island. Fossils are moderately common in the Quatsino and some are very well preserved. These include diverse ammonites and bivalves, in addition to gastropods and a few colonial corals.

Overlying the Quatsino Formation are two quite different rock units of latest Triassic age. On southern Vancouver Island, the Sutton Formation consists of about 100 m of massive, light grey reef limestones composed of broken fragments of shells and colonial corals held together by calcareous algae. On the south side of Cowichan Lake, the best locality of the Sutton, fossils are abundant and well preserved—colonial hexacorals, brachiopods, thick-shelled bivalves, and various ammonites, including heteromorphs. On central and northern Vancouver Island, thin-bedded black limestones and laminated shales of the Parson Bay Formation overlie the Quatsino. The Parson Bay, which is up to 400 m in thickness, was deposited as calcareous muds in deeper water in front of the Sutton reefs. The fossils include paper clams and squashed ammonites, which frequently cover entire bedding surfaces.

PUBLISHED REFERENCES—Triassic

Rocks—Carlisle and Susuki (1965, 1974), Fyles (1955).

Fossils—Carlisle and Susuki (1965), Clapp and Shimer (1911), Tozer (1967, 1979, 1984).

Figure 23 The hexacoral *Elysastrea whiteavesi* (RBCM) and the bivalve *Minetrigonia suttonensis* (GSC 14256). Both from the Sutton Fm., Lake Cowichan.

HEXACORALS

Elysastrea (Figure 23). The Upper Triassic reefs of the Sutton Formation near Cowichan Lake contain large dome-like colonial hexacorals with polygonal corallites defined by strong rims and with depressed centres. Like many other fossils in this formation, the original calcium carbonate shell material of *Elysastrea whiteavesi* has been replaced by silica.

BIVALVES

Minetrigonia (Figure 23). Trigoniid bivalves are moderately common fossils in Mesozoic strata. The Late Triassic *Minetrigonia*, the oldest trigoniid on Vancouver Island, is a typical member of the family. It is triangular in outline with strong radial ribs and complex striated hinge teeth. Silicified specimens of *M. suttonensis* are present in the Sutton Formation at Cowichan Lake. The only trigoniid bivalve in existence today is a single genus living in the waters around Australia. This bivalve, *Neotrigonia*, possesses a large L-shaped foot that is so muscular the bivalve can actually jump above the sand when disturbed.

Figure 24 The bivalves *Halobia alaskana* (GSC 85846), Parson Bay Fm., Nimpkish Lake and *Monotis subcircularis* (GSC 85850), Parson Bay Fm., Neroutsos Inlet.

Monotis (Figure 24). Many shale bedding surfaces of the Parson Bay Formation are covered by massed specimens of thin-shelled bivalves. The mode of life of these paper clams is not well understood. Most likely, they lived byssally attached to floating seaweed, and when they died or the seaweed rotted, the valves drifted down to accumulate on the sea bottom. *Monotis* is the most characteristic of Triassic paper clams. It is obliquely oval in outline with radial ribs and an umbo flanked by small wings. *Monotis* had a remarkably short duration. This genus lasted for only one or two million years in the latest Triassic, but within this brief period, it attained a world-wide distribution. A couple of species occur on Vancouver Island, the most common being *M. subcircularis*.

Halobia (Figure 24). The other common paper clam from the Parson Bay Formation, *Halobia*, is semicircular in outline with a wide hinge line and a sculpture of fine radial grooves and concentric folds. The anterior wing is outlined by a sharp groove. Unlike the superficially similar *Monotis*, *Halobia* lasted through much of the Middle and Late Triassic.

AMMONITES

Figure 25 The ammonites *Paratropites sellai* (GSC 17980) and *Tropites dilleri* (GSC 17986). Both from Quatsino Fm., Open Bay, Quadra Island.

Tropitid ammonites (Figure 25). The Upper Triassic family Tropitidae is well represented in the grey limestones of the Quatsino Formation on central and northern Vancouver Island. The genus *Tropites* includes compact, rounded, rather involute ammonites with low, rounded ribs and a conspicuous tube-like keel. The inner whorl of *Tropites dilleri* has a characteristic barrel shape. *Paratropites sellai* bears deep, sweeping furrows and a strong, rim-like keel. Quatsino limestones exposed at Open Bay on Quadra Island contain particularly well-preserved specimens of tropitid ammonites that have been replaced by silica.

Trachysagenites (Figure 26). Looking more like a circular rasp than an ammonite, *Trachysagenites herbichi* from the Quatsino Formation on Quadra Island is involute with conspicuous spiral sculpture of fine spines arranged in dense rows on the ribs.

Figure 26 The ammonites *Trachysagenites herbichi* (GSC 17981) and *Stantonites* cf. *rugosus* (GSC 17982). Both from Quatsino Fm., Open Bay, Quadra Island.

Stantonites (Figure 26). *Stantonites* cf. *rugosus* from the Quatsino Formation on Quadra Island is an evolute ammonite with coarse, rounded ribs, a pair of prominent rows of tubercles on the lateral edge, and a sharply incised groove on the venter.

Parajuvavites (Figure 27). *Parajuvavites* is a large compressed involute ammonite with faint rounded ribs on the flanks that become more conspicuous toward the venter. The key feature of this ammonite is the last volution of the shell, which becomes eccentric; in effect, it is on its way to becoming a heteromorph. *Parajuvavites* sp. occurs in the Upper Triassic Parson Bay Formation at Kyuquot Sound.

Parson Bay ammonites (Figure 28). Poorly preserved and flattened ammonites are quite common in the black shales of the Parson Bay Formation, but many are so squashed that they cannot be identified. Those that can be identified to the genus level include *Gnomohalorites*, which has rows of rounded tubercles on each rib; *Sagenites*, which bears fine spiral sculpture; and the straight heteromorph *Rhabdoceras*, having fine ribs.

Figure 27 The ammonite *Parajuvavites* sp. (GSC 14306), Parson Bay Fm., Kyuquot Sound.

Figure 28 Shale surface with three ammonites—the knobby *Gnomohalorites*, the finely lined *Sagenites*, and the small straight heteromorph *Rhabdoceras* (open arrow) (GSC 28416), Parson Bay Fm., Malksope Inlet.

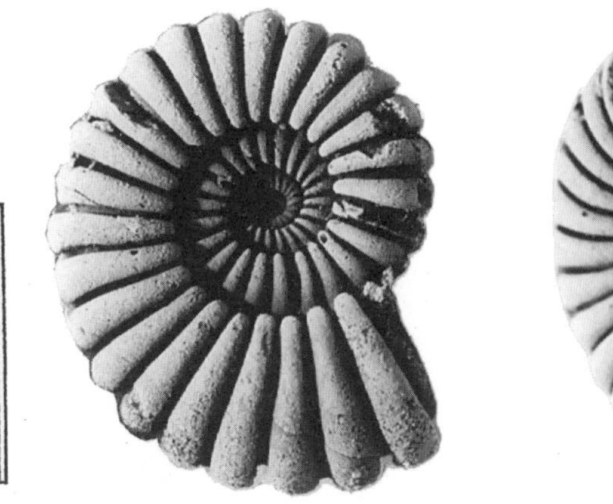

Figure 29 The ammonite *Sympolycyclus gunningi* (GSC 14236), Quatsino Fm., Open Bay, Quadra Island.

Figure 30 The ammonites *Cycloceltites* sp. (GSC 17015), *Choristoceras suttonense* (GSC 32326), and the heteromorph ammonites *Paracochloceras canaliculatum* (GSC 32329) and *Rhabdoceras suessi* (GSC 32317). All from Sutton Fm., Lake Cowichan (with the exception of *P. canaliculatum*, which is from Parson Bay Fm., Quatsino Sound).

Choristoceratid ammonites (Figures 29 and 30). Rocks of latest Triassic age on Vancouver Island contain excellent examples of choristoceratids, an unusual group of small ammonites characterized by serpenticone or heteromorphic coiling and a sculpture of simple ribbing. *Sympolycyclus gunningi* is an exquisite serpenticone ammonite with swollen ribs separated by narrow constrictions. The spare elegance of this ammonite and its classical dimensions would not be out of place on an Art Deco frieze. *Cycloceltites* sp. is another serpenticone with irregular and well-spaced ribs. The inner whorls of *Choristoceras suttonense* are serpenticone, but the outer whorl of this heteromorph is detached into a loose spiral. *Rhabdoceras suessi* starts as a few serpenticone whorls before it becomes a closely ribbed straight rod. *Paracochloceras canaliculatum*, the most unusual of all Triassic heteromorphs and also one of the rarest, is a high sinistrally coiled turret with low slanted ribs. These Late Triassic heteromorphs are similar to the heteromorphs that appear in the Late Cretaceous, but they are entirely unrelated in that each was derived from a different type of planispiral ammonite.

FISHES

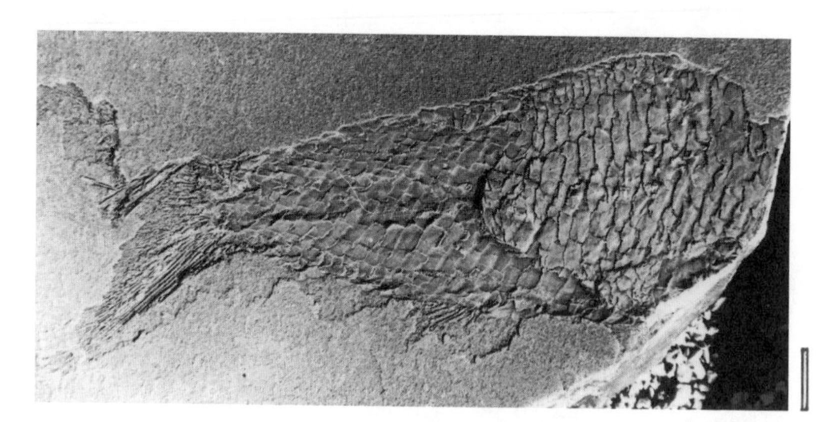

Figure 31 The ray-finned fish *Australosomus* sp. (Port Hardy Museum). Parson Bay Fm., Keogh River.

Australosomus (Figure 31). It is not complete and it still needs to be prepared, but this 15-cm-long specimen is extremely important because it is the first articulated fossil fish discovered on Vancouver Island. Moreover, it is the first fish known from Triassic rocks on the island. The deeply cleft tail fin, the posteriorly located anal fin, and the rectangular interlocking body scales that are three or four times as high as wide, identifies this specimen as *Australosomus*, a genus previously known from Lower Triassic rocks of Greenland, Africa, Madagascar, and central British Columbia. *Australosomus* is one of many groups of ray-finned fishes that arose in the Triassic after the great extinction at the end of the Permian but that did not survive into the Jurassic. This specimen was collected from black, finely laminated limestones of the Parson Bay Formation (Upper Triassic) at the Island Highway bridge across the Keogh River between Port McNeill and Port Hardy.

JURASSIC

West coast collectors rarely see Jurassic fossils because almost all Jurassic sedimentary rocks on Vancouver Island are exposed along the inaccessible and roadless northwestern coast between Nootka Sound

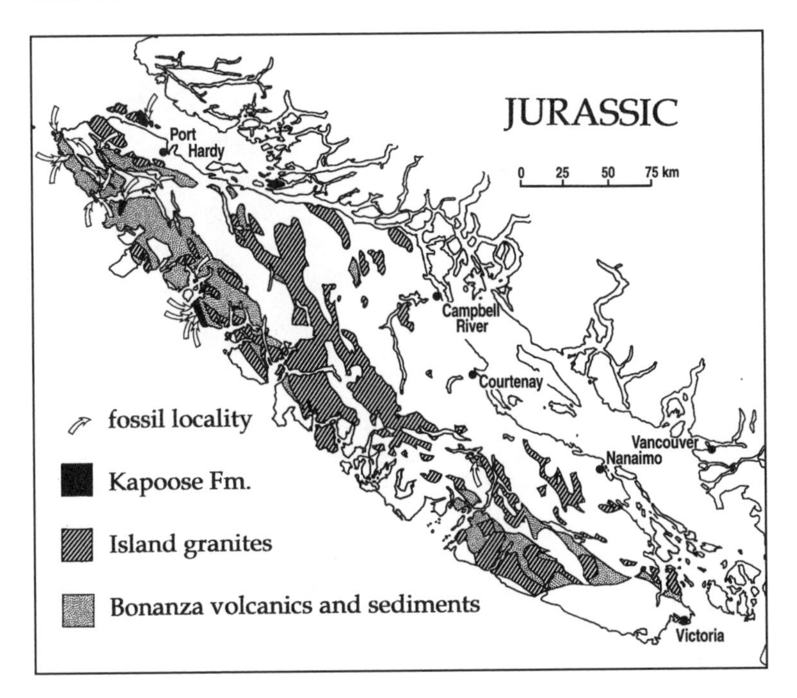

Figure 32 Distribution of Jurassic rocks on Vancouver Island and fossil localities.

and Cape Scott (Figure 32). Well-preserved Jurassic ammonites have been collected from rocks exposed at Quatsino Sound, Esperanza Inlet, Kyuquot Sound, and Tahsis Inlet. However, these localities are better known to west coast naturalists as the focal points in the struggle to save the few remaining large stands of ancient first-growth forest from logging.

The oldest Jurassic rocks on Vancouver Island are thick accumulations of basalts (once lavas) and tuffs (at one time volcanic ashes), alternating with siltstones and gritty sandstones. These rocks of the Bonanza Group, thousands of metres thick, are exposed for the full length of the island from Saanich Inlet to Cape Scott. They weather in characteristic red, maroon, or lavender colours due to the presence of iron oxides. The shales, siltstones, and sandstones of the Bonanza contain infrequent fossils of Early Jurassic age (the best are from the Memekay River area south of Sayward) and Middle Jurassic age (known only in the Koksilah River area south of Duncan). These fossils are primarily ammonites and bivalves.

Extensive areas of western Vancouver Island are underlain by light-coloured granitic rocks assigned to the Island Granites—igneous rocks that intruded Early Jurassic and older rocks during the mid-Jurassic. Overlying the Island Granites are brown concretionary siltstones and limestones of the Kapoose Formation, exposed along the west coast between the head of Kyuquot Channel and Esperanza Inlet. The Kapoose Formation, the youngest Jurassic unit on Vancouver Island, contains well-preserved ammonites, nautiloids, bivalves, and belemnites at many localities.

Global biogeographic patterns indicate that Wrangellia had reached the northern mid-latitudes by the Jurassic. The types of ammonites present suggest that this terrane was straddling the boundary between a north temperate province and a tropical province during the Jurassic.

PUBLISHED REFERENCES—Jurassic

Rocks—Carson (1973), Jeletzky (1976), Muller, Cameron, and Northcote (1981).

Fossils—Frebold (1964), Frebold and Tipper (1970), Jeletzky (1965), Smith and Tipper (1986).

BIVALVES

Figure 33 The bivalves *Gresslya* sp. (VIPM 006), *Vaugonia* cf. *coatsi* (VIPM 007), and *Opisoma* sp. (VIPM 008). All from Bonanza Group, south of Sproat Lake.

Gresslya (Figure 33). *Gresslya* is an oval clam from the Bonanza Group south of Sproat Lake, with a conspicuous umbo and low, rounded growth lines. It was undoubtedly a shallow burrower.

Vaugonia (Figure 33). Each valve of this oblong trigoniid bivalve consists of a lower triangular area marked by coarse ridges and a finely striated upper area. The photographed specimens show the external and an internal mold of two right valves. All trigoniid bivalves were (and are) shallow burrowers. *Vaugonia* cf. *coatsi* occurs in siltstones among the Bonanza volcanics at Sproat Lake.

Opisoma (Figure 33). Shaped like a duck's foot, *Opisoma* sp. has three or four ribs sweeping back from the umbo and ending as blunt prongs. This bivalve probably reclined on top of the mud stabilized by the prongs, which acted as outriggers.

Pinna (Figure 37). Living *Pinna*—pen shells—are wedge-shaped, thin-shelled bivalves that live in shallow water; they carry on their lives in an upright position, partially buried in sand, and are held in place by a byssus. In the Mediterranean region, these byssal filaments are sometimes spun into threads and woven into a fine fabric called "cloth of gold." Fossil *Pinna*, which look just about the same as living pen shells, must have lived in the same way. The large Jurassic specimen

Figure 34 The bivalve *Pholadomya* sp. (VIPM 149) and the ammonite *Paltechioceras* sp. (VIPM 150). Both from Bonanza Group, Memekay River.

of *Pinna* sp. from the Kapoose Formation is high and pyramidal, with arched growth lines on the sides.

Pleuromya (Figure 38). This common Jurassic bivalve is inflated and triangular in outline, with coarse concentric undulations. *Pleuromya* cf. *alduini* occurs in the Kapoose Formation at Kyuquot Sound.

Pholadomya (Figure 34). *Pholadomya* is a thin-shelled bivalve with a rounded front margin and a permanent opening for the siphons at the rear. Radiating ribs extend across the middle of the shell. *Pholadomya* sp. is a common fossil in the Bonanza Group at the Memekay River site south of Sayward. The genus is now living as a shallow burrower in fairly deep water in the Caribbean.

Retroceramus (Figure 36). On Vancouver Island, as elsewhere, the vast majority of inoceramid bivalves occurs in Cretaceous rocks, but some genera are found in rocks as old as Permian. *Retroceramus* cf. *porrectus* is an elongate, mussel-shaped inoceramid with a pointed umbo that is characterized by broad, concentric corrugations of sharp-edged folds. It occurs in a dark siltstone unit of Middle Jurassic age from the Bonanza volcanics exposed on the Koksilah River.

AMMONITES

Figure 35 The ammonites *Oxynoticeras* sp. (VIPM 009) and *Paltechioceras* sp. (VIPM 010, 146). All from Bonanza Group, south of Sproat Lake.

Oxynoticeras and ***Paltechioceras*** (Figures 34 and 35). One particular fossil collection from siltstones among volcanics of the Bonanza Group south of Sproat Lake includes two ammonites that occupy opposite ends of the ammonite morphologic spectrum. One is involute and smooth with a simple keel (an oxycone); the other is evolute and ribbed with a complex keel (a serpenticone). The oxycone *Oxynoticeras* sp. looks like a sharp-edged discus—it has gentle, almost imperceptible undulations on the flanks, and a razor-sharp keel. The serpenticone *Paltechioceras* sp., coiled like a rope, bears coarse transverse ribs that are interrupted by three low ridges on the keel.

Stephanoceras (Figure 36). This characteristic fossil demonstrates that Middle Jurassic sediments do occur within the Bonanza volcanics

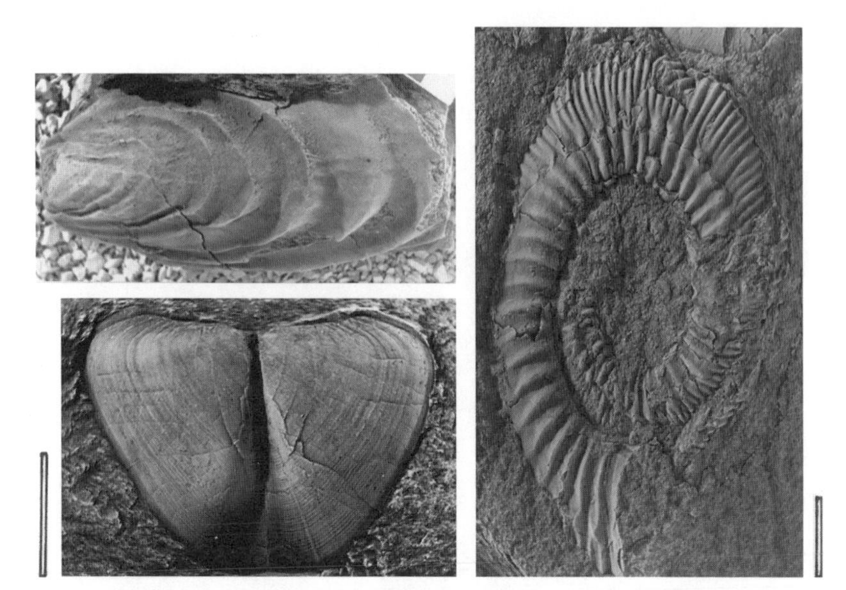

Figure 36 The bivalve *Retroceramus* sp. (C. Ruttan Collection), the ammonite *Stephanoceras* sp. (GSC 117612), and an ammonite lower jaw (C. Ruttan Collection). All from Bonanza Group, Koksilah River.

exposed on the Koksilah River, west of Shawnigan Lake and south of Duncan on southern Vancouver Island. *Stephanoceras* is an evolute ammonite studded by a conspicuous row of nodes at mid-whorl. Below each of these nodes extends a single rib; above it, two and occasionally three smaller ribs continue across the venter. *Stephanoceras* had a world-wide distribution for a brief time in the early part of the Middle Jurassic. It has not previously been known on the west coast.

Ammonite lower jaw (Figure 36). The lower jaws (or aptychi) of ammonites are moderately common in Upper Cretaceous rocks of Vancouver Island. These fossils are U-shaped and wrinkled. They must have been robust because they are strengthened by deposits of calcite (see Figure 88). The jaws of older ammonites are much more rare. These aptychi are composed of flexible organic material, but are not thickened by calcite. They are preserved spread out, butterfly-like. An exceptionally well-preserved Middle Jurassic specimen from the Bonanza volcanics at Koksilah River is inscribed by very fine growth lines.

Figure 37 The bivalve *Pinna* sp. (GSC 103317) and the ammonite *Phylloceras* cf. *apenninicum* (GSC 103318). Both from Kapoose Fm., Kyuquot Sound.

Phylloceras (Figure 37). Evolution and extinction of ammonite families proceeded at a hectic pace during the Mesozoic. Only a few families lasted as long as a system—the phylloceratids are alone in persisting from the Triassic to the Late Cretaceous. Tim Tozer, the foremost authority on Triassic ammonites, believes that all Jurassic and Cretaceous ammonites were derived from a few phylloceratids that escaped the mass extinction at the end of the Triassic. The Jurassic *Phylloceras* cf. *apenninicum* from the Kapoose Formation at Kyuquot Sound is almost identical to the Late Cretaceous *Neophylloceras ramosum* from the Nanaimo area (see Figure 90). *Phylloceras* is an involute compressed ammonite covered by fine dense lirae and low corrugations.

Figure 38 The belemnite *Pachyteuthis* sp. (GSC 103319), the ammonite *Cardioceras* cf. *alphacordatum* (GSC 103320), and the bivalve *Pleuromya* cf. *alduini* (GSC 103321). All from Kapoose Fm., Kyuquot Sound.

Figure 39 Shale slab crowded with the ammonite *Cardioceras* sp. (GSC 103322), Kapoose Fm., Kyuquot Sound.

Cardioceras (Figures 38 and 39). This distinctive compressed ammonite has well-developed curved primary ribs with smaller secondary ribs and, most characteristic of all, bears a keel resembling braided rope. At least two species occur in separate collections from the Kapoose Formation at Kyuquot Sound. *Cardioceras* sp. has simple primary ribs; C. cf. *alphacordatum* has swollen primary ribs.

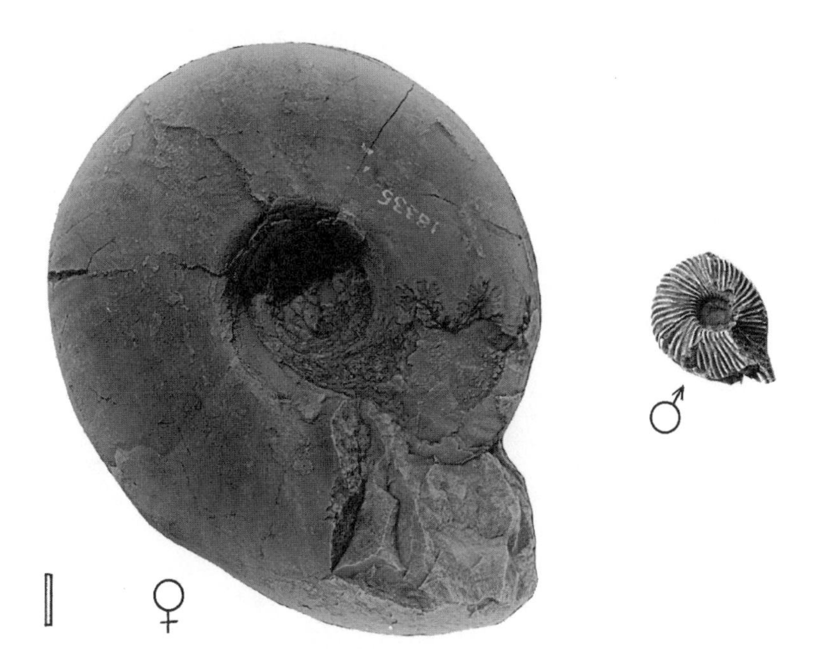

Figure 40 The ammonite *Cadoceras* sp. (GSC 103325, 103326), probable female and male morphs, Kapoose Fm., Kyuquot Sound.

Cadoceras (Figure 40). In living animals, invertebrates as well as vertebrates, the two sexes are frequently expressed by size and shape differences. Among invertebrate groups, females are commonly larger than males. In mammals, the males are frequently slightly larger, but in many fishes and birds, the males are smaller than the females. Sexual dimorphism can also be determined in fossils but, of course, such evidence can never be conclusive. More work has been done on sexual dimorphism of ammonites than on any other group of fossils.

A large collection from the Kapoose Formation at Kyuquot Sound includes large and small versions of *Cadoceras* sp. These probably represent separate sexes. *Cadoceras* is a moderately involute ammonite with dense sharp ribbing on immature whorls. This ribbing gradually diminishes so that mature whorls are almost smooth. Some specimens apparently ceased growth after a few whorls (so-called microconchs); others continued growing and adding whorls (macroconchs). The macroconchs were probably females; the microconchs males.

NAUTILOIDS

Figure 41 The nautiloid *Cenoceras* cf. *intermedius* (GSC 103327), Kapoose Fm., Tatchu Point, Kyuquot Sound.

Cenoceras (Figure 41). Nautiloids are moderately common in Jurassic rocks on Vancouver Island. A well-preserved specimen occurs in the Kapoose Formation in the Kyuquot Sound area. *Cenoceras* cf. *intermedius* is involute with nearly straight sutures and a flattened venter, giving a rectangular outline to the body chamber. The siphuncle appears to occupy a central position.

BELEMNITES

Pachyteuthis (Figure 38). Belemnites are an extinct group of squid-like cephalopods that left a good fossil record of cigar-shaped shells composed of solid radiating crystals of calcite. The solid shell envelops a central septate cone—disclosing its affiliation with other chambered cephalopods, such as nautiloids. Belemnites extend from the Carboniferous to the early Cenozoic, but the vast majority of specimens are found in Jurassic and Cretaceous rocks. As the name implies, *Pachyteuthis* sp. from the Kapoose Formation at Kyuquot Sound is a robust belemnite.

CRETACEOUS

The ferry from Tsawwassen to Swartz Bay on Vancouver Island has to thread the needle known as Active Pass, between Galiano Island and Mayne Island. The manoeuvres are difficult, involving two right-angle turns in a narrow channel often crowded with small boats and swept by strong tidal currents. First-time travellers, seeing the sandstone promontory rise immediately in front of the ferry, may despair of ever reaching Vancouver Island. But somehow the captain manages the sharp turns and the ferry pushes into the open Trincomali Channel inside the islands.

Active Pass is one of the few gaps in an otherwise solid barrier of upfolded Cretaceous strata that rises out of the water to form the chain of outer Gulf Islands. Galiano and Mayne islands are made of alternating sandstone and shale units that are tilted; the resistant sandstones form the headlands that extend into Active Pass, and the recessive shales weather back into bays and coves.

Differential weathering of Upper Cretaceous sandstone and shale units at the foreshore has resulted in a vista that has almost become iconographic of the Gulf Islands—a sailboat at anchor in a tranquil cove flanked by rounded sandstone cliffs overhung by gnarled arbutus trees.

Cretaceous rocks are exposed in two separate regions on Vancouver Island (Figure 42). Lower Cretaceous sandstones of the

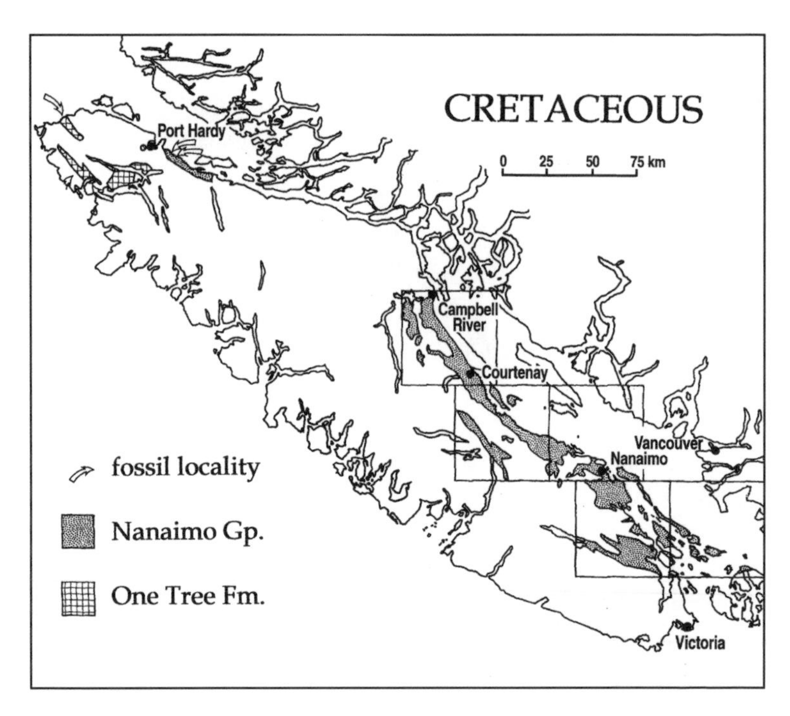

Figure 42 Distribution of Cretaceous rocks on Vancouver Island and fossil localities. The detailed maps of Upper Cretaceous formations are outlined (see Figures 45–49).

One Tree Formation crop out near the inaccessible northern tip of the island. Upper Cretaceous shales, sandstones, and conglomerates of the Nanaimo Group underlie most of the flat populated belt along the east coast and virtually all of the Gulf Islands. In the context of terrane tectonics, the One Tree Formation was deposited prior to the collision of Wrangellia with Laurentia, and the Nanaimo Group was deposited just after this collision.

The Upper Cretaceous sedimentary rocks were deposited in two basins on Vancouver Island and the Gulf Islands—the Comox Basin located between Campbell River and Parksville and on Denman and Hornby islands, and the Nanaimo Basin, which extends from Nanoose Bay south to the Saanich Peninsula and across the southern Gulf Islands to Orcas and Sucia islands.

In both basins, shale formations alternate with sandstone/conglomerate formations (see Figure 9). In the Comox Basin, the Nanaimo Group starts with the Comox Formation (sandstone), followed by Trent River (shale), Denman (sandstone), Lambert (shale), Geoffrey (conglomerate), Spray (shale), and Hornby (conglomerate) formations. Here, this stack of formations is about 2 km thick and gently inclined eastward into the Strait of Georgia. In the Nanaimo Basin, the Nanaimo Group starts with the Benson Formation (conglomerate), followed by Haslam (shale), Extension (conglomerate), Pender (shale), Protection (sandstone), Cedar District (shale), De Courcy (sandstone), Northumberland (shale), Galiano (conglomerate), Mayne (shale), and Gabriola (sandstone) formations. This succession is twice as thick as in the Comox Basin and, furthermore, folded and faulted. This deformation was a result of the collision with two small terranes that rammed under southern Vancouver Island during the Paleogene about 40 million years ago.

With the exception of the Comox, Benson, Extension, and Pender formations, which were deposited partly in non-marine swampy or marginal marine lagoonal environments, all other Upper Cretaceous units of the Nanaimo Group were deposited in marine settings, and mostly at depths between 500 and 1,000 m. At such depths, mud predominates but, because these Late Cretaceous basins on Vancouver Island were narrow and surrounded by uplands on both sides, a lot of sand and gravel from lagoons and beaches was introduced by earthquake-triggered slumps through submarine channels and fans. This means that the shale formations, which contain well-preserved fossils, accumulated rather slowly, while the sandstone and conglomerate formations, which generally lack fossils, were deposited quickly.

The Cretaceous succession on Vancouver Island attracted a lot of attention in the late 1800s because these rocks contained the only commercially viable coal on the entire west coast of North America. Coal, of course, was a strategic energy source during the age of steam. Soon after British Columbia joined Confederation in 1871, James Richardson of the Geological Survey of Canada set out to investigate the coal deposits. Richardson was a farmer from the eastern townships of Quebec who learned his trade as a field geologist in the 1840s as assistant to William Logan, the first director of the GSC. He did excellent independent geological mapping on Anticosti Island in Quebec and in western Newfoundland before he tackled the west coast. Richardson worked out the Cretaceous geology of Vancouver Island by foot and by boat and he collected numerous fossils that were later described by his colleague J.F. Whiteaves, but he found no other major coal deposits.

Coal mines had opened in Nanaimo in the 1850s, and in Cumberland in the 1880s. At their height of activity during the late 1920s, more than three thousand workers toiled in the mines on Vancouver Island. Not much remains of these mines, but well-preserved Cretaceous plant fossils can still be collected from the old mine dumps.

The well-known mass extinction that terminated the dinosaurs, plesiosaurs, ammonites, and many other animal and plant groups at the end of the Cretaceous, has been widely attributed to the catastrophic impact of a large meteorite. The dust and soot thrown up by this impact, which probably occurred on the Yucatan Peninsula, resulted in the near-global deposition of a thin layer of clay rich in the element iridium. This dramatic extinction explanation has proved popular among both scientists and the general public. However, growing evidence now indicates that extinctions of both marine and terrestrial animals and plants were well underway by the time the iridium-rich layer was laid down 65 million years ago. These extinctions probably occurred over a prolonged period and thus cannot be attributed to a single cause. They seem to be the biotic response to massive environmental deterioration resulting from a complex series of earth-bound events—including a sea-level drop, a decrease in temperature, changes in ocean circulation, and oxygen depletion in the oceans. The impact of the extra-terrestrial body only sharpened a gradual decline in biological diversity. But the debate still rages in geological and paleontological journals: Did the Cretaceous end with a bang, or a whimper?

PUBLISHED REFERENCES—Cretaceous

Rocks—Buckham (1947), England (1989), England and Calon (1991), England and Hiscott (1992), McLellan (1927), Muller and Jeletzky (1970), Mustard (1994), Ward (1978).

Fossils—Anderson (1958), Bell (1957), Feldmann and McPherson (1980), Haggart (1989, 1991, 1996), Haggart and

Ward (1989), Jeletzky (1965), Jones (1963), Ludvigsen (1996), Nicholls (1992), Rathbum (1926), Richards (1975), Saul (1978), Usher (1952), Ward (1976, 1978, 1985), Ward and Mallory (1977), Whiteaves (1879, 1895, 1903).

LOWER CRETACEOUS

Virtually all of the Cretaceous fossils dealt with in this section are of Late Cretaceous age and from the eastern part of Vancouver Island and the Gulf Islands. A few Lower Cretaceous fossils from the north coast of Vancouver Island are illustrated in Figures 43 and 44.

BIVALVES

Figure 43 The bivalve *Buchia pacifica* (RBCM), One Tree Fm., Kyuquot Sound.

Buchia (Figures 43 and 44). Many shallow-water sandstones of Late Jurassic and Early Cretaceous age in British Columbia contain little else but massed specimens of a single species of *Buchia*. This bivalve has an incurved left valve overhanging the right valve. Both valves bear conspicuous growth lines. The weight of the right valve kept the

Figure 44 The ammonite *Tollia* cf. *simplex* (GSC 16627), the bivalve *Buchia okensis* (GSC 17471), and the belemnite *Cylindroteuthis* cf. *obeliscoides* (GSC 16582). All from One Tree Fm., Clark Island, Kyuquot Sound.

bivalve stable in the sand, where it probably lived as a filter-feeder. A prolific aggregation from the One Tree Formation at Kyuquot Sound consists solely of swollen valves of B. *pacifica*. Different species occur at other fossil localities in the same formation—the much flatter B. *okensis*, for example.

AMMONITES

Tollia (Figure 44). This small, compressed, moderately evolute ammonite bears sharp ribs that split near the venter. *Tollia* cf. *simplex* is a rare fossil in the One Tree Formation on Clark Island off Kyuquot Sound.

BELEMNITES

Cylindroteuthis (Figure 44). This belemnite from the One Tree Formation consists of a long slender shell with a pointed apex. The resemblance to an Egyptian obelisk is emphasized by the name *Cylindroteuthis* cf.

obeliscoides. The fossils of the One Tree Formation on northern Vancouver Island are particularly associated with George Jeletzky—the Mesozoic paleontologist to the Geological Survey of Canada for nearly forty years. Jeletzky was born in Russia and educated at the University of Kiev, graduating in 1941 just before the invasion of Russia. He managed to keep possession of a precious collection of Cretaceous belemnites while he and his family were buffeted across eastern Europe by the ebb and flow of the Soviet and German armies. He came to Canada in 1948 to work for the GSC and was assigned a field area in northern Vancouver Island. Over the years he became the world's foremost authority on belemnites—these deceptively simple cigar-shaped fossils. George Jeletzky, who died in 1988, is responsible for most of our knowledge about Mesozoic geology and paleontology of Vancouver Island.

UPPER CRETACEOUS

The components of the rich fossil faunas and floras of the Upper Cretaceous Nanaimo Group are illustrated in Figures 52 to 145.

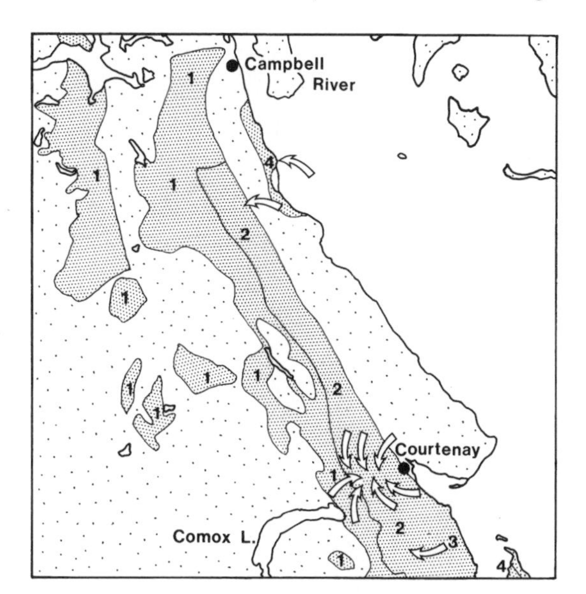

Figure 45 Upper Cretaceous geology in the Campbell River-Courtenay area and fossil localities (see legend on Figure 47).

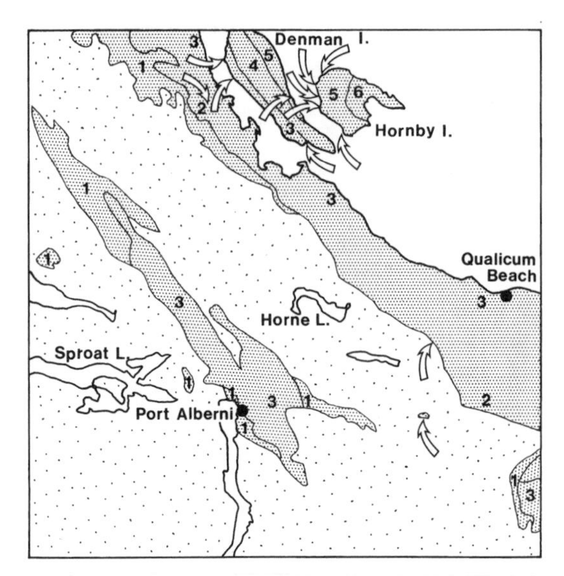

Figure 46 Upper Cretaceous geology in the Port Alberni-Qualicum Beach area and fossil localities (see legend on Figure 47).

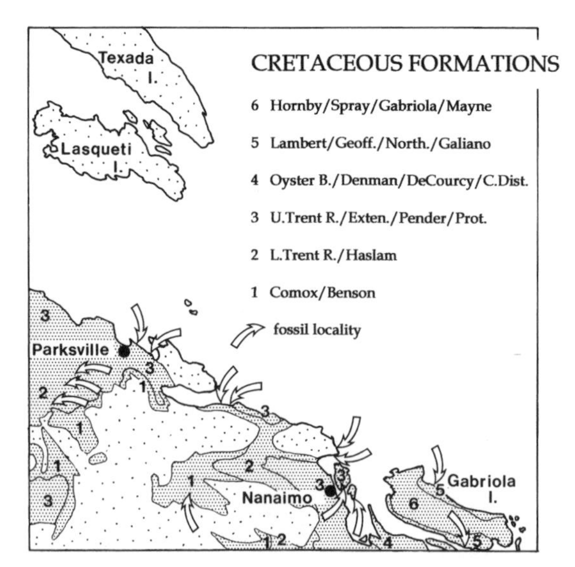

Figure 47 Upper Cretaceous geology in the Parksville-Nanaimo area and fossil localities.

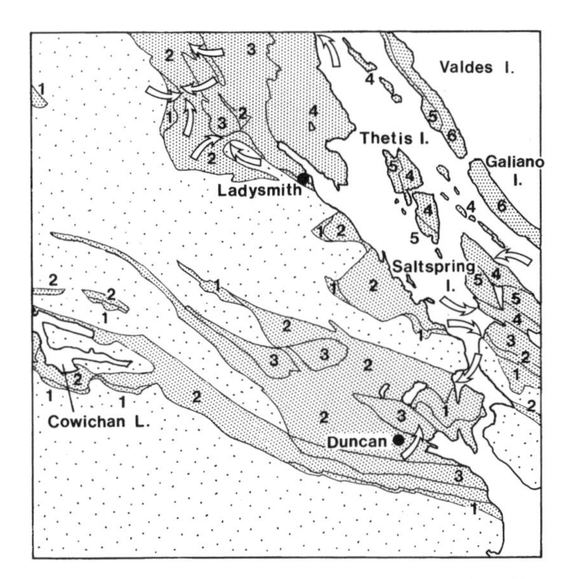

Figure 48 Upper Cretaceous geology in the Ladysmith-Duncan area and fossil localities (see legend on Figure 47).

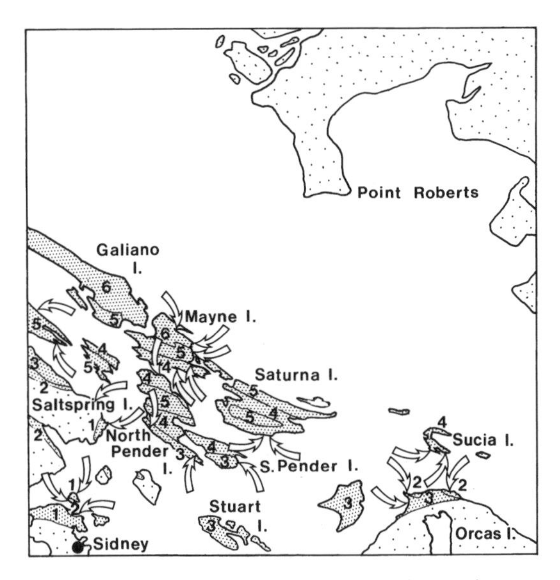

Figure 49 Upper Cretaceous geology in the southern Gulf Islands-San Juan Islands area and fossil localities (see legend on Figure 47).

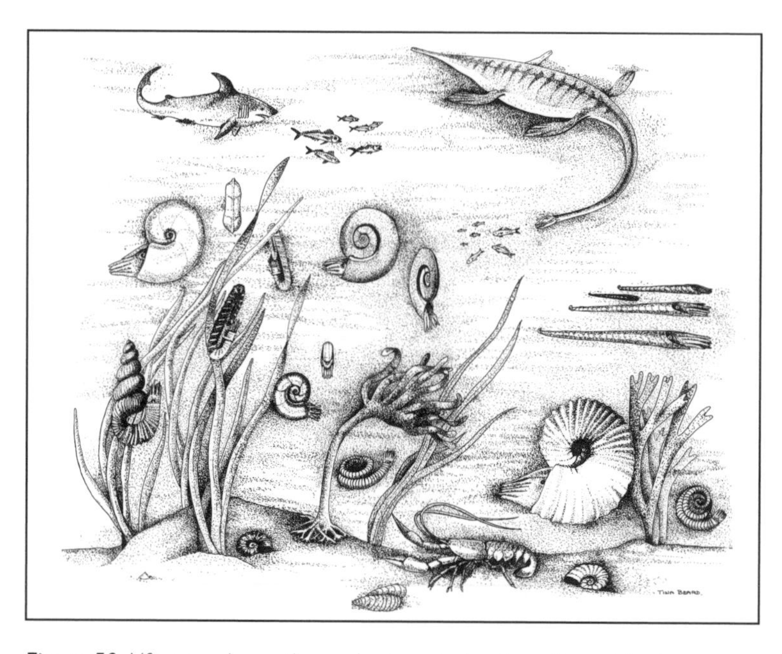

Figure 50 Life on and near the sea bottom during deposition of the lower Trent River Formation exposed on the Puntledge and Browns rivers. A great variety of planispiral and heteromorph ammonites hunt for small fish above the bottom and among the seaweed. In the background, an elasmosaur is closing in on a school of fish. Different gastropods, bivalves, and lobsters populate the muddy bottom (artist Tina Beard).

BIVALVES

Glycymerita (Figure 52). These bivalves are oval in outline with inflated, finely ribbed valves and large teeth that are arranged in an arc. The ligamental area below the umbo is scored by fine chevron furrows. The thick shell consists of two layers—an inner lamellar layer and an outer prismatic layer, well illustrated in the polished cross-section. *Glycymeris*, a closely related genus, commonly called bittersweet shells, still lives in the waters around Vancouver Island, where it is a shallow burrower in mud and silt. *Glycymerita veatchii* is a common fossil bivalve in the Haslam Formation at Brannan Lake and in the Lambert Formation on Hornby Island.

Figure 51 Life on and near the sea bottom during deposition of the Lambert Formation exposed at Collishaw Point on Hornby Island. Heteromorph and planispiral ammonites compete for prey while a pair of camouflaged cuttlefish lurk in the seaweed. A great variety of animals crowd the bottom—including large clams, spiny lobsters, crabs, and gastropods. High above the bottom, a mosasaur is hunting a group of ammonites while sharks pursue a school of fish (artist Tina Beard).

Yaadia (Figure 53). An oddly named fossil, *Yaadia* is a large trigoniid bivalve with a lumpy sculpture of irregular knobs arranged in oblique rows. Louella Saul, a Los Angeles paleontologist who has written the definitive monograph on these bivalves, commented that "the sculpture of *Yaadia* is typically untidy." The upper flat area carries a row of fine transverse ribs flanking a long swollen ligament. Like *Neotrigonia*, the sole living trigoniid, *Yaadia* was a shallow burrower. *Y. tryoniana* occurs in the Haslam Formation and in the lower Trent River Formation at Northwest Bay.

Pterotrigonia (Figure 54). *Pterotrigonia* is a markedly triangular trigoniid bivalve having high vertical ribs and an inwardly twisted umbo. The upper area is concave with strong transverse ribs that are demarcated

Figure 52 Three specimens of the bivalve *Glycymerita veatchii* (VIPM 011, 012, 013), Haslam Fm., Brannan Lake.

Figure 53 The bivalve *Yaadia tryoniana* (VIPM 014), Haslam Fm., Brannan Lake.

Figure 54 The bivalve *Pterotrigonia evansana* (VIPM 015, 016), Trent River Fm., Northwest Bay and Haslam Fm., Brannan Lake.

by a raised smooth belt. Even on small shell fragments, the sculpture of *Pterotrigonia* is unmistakable. Although *P. evansana* is frequently found in the Haslam Formation, it is particularly common in sandstones of the lower Trent River Formation and in the Comox Formation near Mount Washington, northwest of Courtenay. Another species, *P. klamathonia*, from the Benson Formation on Sidney Island, bears small nodes on the high ribs.

Opis (Figure 55). This odd shell is one of the rarest bivalves in Cretaceous rocks of Vancouver Island or the Gulf Islands. *Opis* is an oblong, heavy-shelled bivalve with fine growth lines. The shell interior is equally divided into a deep, circular body cavity and a triangular hinge area composed of a pair of massive, plate-like teeth separated by deep sockets. The flanks of the teeth are scored by fine grooves. *Opis vancouverensis* occurs at a single locality in the upper Trent River Formation on Denman Island. A different species occurs in the Haslam Formation at Brannan Lake.

Odontogryphaea (Figure 56). The oysters that are abundant today in intertidal areas along the northern Strait of Georgia belong to the introduced species *Crassostrea gigas*, the Japanese oyster, which has almost entirely displaced the native oyster *Ostrea lurida*. Cretaceous fossil oysters differ from both of these species in being broadly triangular with extremely thick shells that have finely crenulated margins

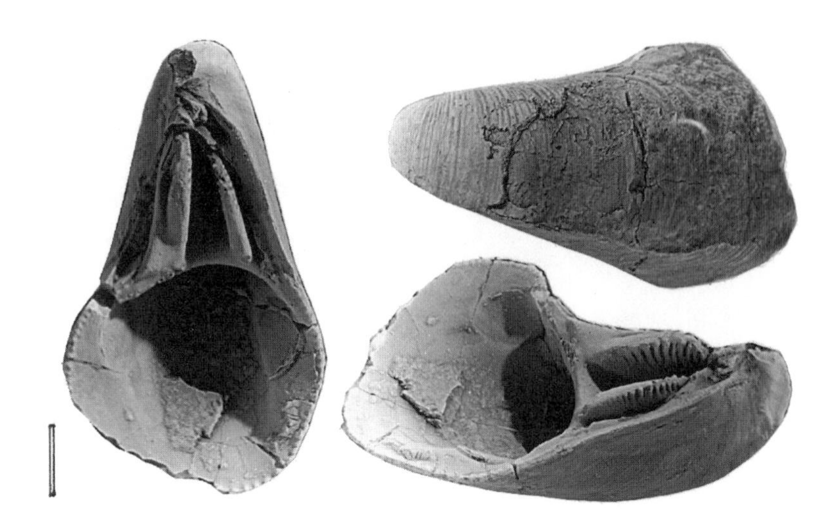

Figure 55 The bivalve *Opis vancouverensis* (VIPM 151), Trent River Fm., west side of Denman Island.

and a vague fold on the left valve. These oysters from the Haslam Formation are best assigned to *Odontogryphaea* sp.; like all oysters, it has a single muscle and lacks hinge teeth.

Lithophaga (Figure 56). A living date mussel can bore into thick calcite shells or carbonate rock by using an organic acid it secretes. The periostracum (an enveloping organic sheath) protects its own calcite shell from similar dissolution. Fossil *Lithophaga* sp. have been discovered in crypts etched into the massive left valves of the oyster *Odontogryphaea* sp. from the Haslam Formation.

Inoceramus (Figure 57). At virtually every outcrop of Upper Cretaceous shales on Vancouver Island and the Gulf Islands, fragments of large flat bivalve shells are the most conspicuous fossils. These impressive dinner-plate-sized valves belong to *Inoceramus*—Greek for "fibrous vessel," so named because their centimetre-thick shell is composed of fine, asbestos-like fibres of calcite. This inoceramid bivalve is concentrically corrugated with superimposed fine growth lines and, on occasion, with fine radial ribs. Extinct since the Cretaceous, these bivalves must have lain flat on the surface of the mud where they may have

Figure 56 The bivalves *Odontogryphaea* sp. (VIPM 017) and *Lithophaga* sp. (VIPM 018). Both from Haslam Fm., Brannan Lake.

Figure 57 The bivalve *Inoceramus vancouverensis* (VIPM 019), Lambert Fm., Collishaw Point, Hornby Island.

been filter-feeders. They reached considerable dimensions—specimens a metre across have been reported.

Because they lived in deep dark water, low in oxygen but high in hydrogen sulphide, it is possible that the fleshy mantle of *Inoceramus*, like modern deep-sea organisms living next to submarine volcanic vents, contained chemosynthetic symbiotic bacteria that produced sugars through fermentation. However, this possibility remains speculative. A number of species of *Inoceramus* occur in the Upper Cretaceous—the best known being *I. vancouverensis*, which is particularly common in the Lambert Formation on Hornby Island.

Sphenoceramus (Figures 58 and 59). This inoceramid bivalve is long and triangular in outline with a pointed umbo. The sculpture is highly characteristic, consisting of concentric folds intersected by strong, superimposed oblique and radial ribs. *Sphenoceramus schmidti* occurs in the Trent River Formation, for example at Northwest Bay. *S. elegans* is moderately common in the Haslam Formation of the Nanaimo area. *S. naumanni* is abundant in the lower Trent River Formation shales exposed on the Puntledge, Browns, and Trent rivers. This species bears dense regular concentric undulations, but it lacks any oblique or radial ribs.

Spondylus (Figure 59). Living species of *Spondylus* are eagerly sought after by shell collectors because of their brilliant hues of orange, red, and violet. Known as "spiny oysters" because they cement to hard objects like true oysters, these bivalves are, however, of the scallop family. *S. spinosus* is a rare fossil in the uppermost Trent River Formation exposed on the west side of Denman Island. It carries low radial ribs that are interrupted by regular rows of spines. Each valve bears a pair of robust teeth and corresponding sockets.

Willamactra (Figure 60). In modern seas, mactrids are large, flat, smooth surf clams that have a permanent gape of the valves for fused siphons to facilitate shallow burrowing. *Willamactra truncata* has a broad oval outline and a sculpture of closely spaced growth lines. *W. suciensis* is somewhat more inflated with more conspicuous growth lines. Both species occur in the Haslam Formation at Brannan Lake.

Cymbophora (Figure 61). This surf clam is broadly triangular in outline with a prominent crease along its posterior edge. The sculpture of fine,

Figure 58 The bivalves *Sphenoceramus schmidti* (VIPM 020, 021), Trent River Fm., Northwest Bay and *Sphenoceramus elegans* (VIPM 022), 023), Haslam Fm., Brannan Lake.

Figure 59 The bivalves *Sphenoceramus naumanni* (VIPM 160), Trent River Fm., Browns River and *Spondylus spinosus* (VIPM 159), Trent River Fm., west side of Denman Island.

Figure 60 The bivalves *Willamactra truncata* (VIPM 024) and *Willamactra suciensis* (VIPM 025). Both from Haslam Fm., Brannan Lake.

Figure 61 The bivalves *Cymbophora warrenana* (CDM 008, 009) and *Mytiloides duplicostatus* (CDM 010). All from Trent River Fm., Oyster River.

closely spaced concentric lines is characteristic. Like all modern surf clams, *Cymbophora* was probably a shallow burrower in sand. *C. warrenana* is the only Cretaceous species from Vancouver Island that can definitely be assigned to this genus.

Mytiloides (Figure 61). *Mytiloides* is a triangular inoceramid bivalve with small wings flanking a pointed umbo. The sharp concentric undulations are conspicuously coupled in pairs. M. *duplicostatus* is found in sandstones of the Trent River Formation along the Oyster River west of Campbell River.

Teredina (Figure 62). During the age of wooden ships, shipworms were the bane of the maritime industry. Even now, they do considerable damage to coastal wooden structures, such as docks and pilings, and to wooden barges. And yet, as naturalists, we can't help delighting in the sheer ingenuity of these small bivalves, which, during the Mesozoic, discovered the knack of boring into wood and, with the aid of intestinal bacteria, even the ability to digest cellulose. *Teredina* consists of two short hooked valves with a pair of longitudinal furrows, each valve scored by fine transverse ridges. Excavation of the borings was achieved by the filing action of these ridges as the valves were rocked back and forth. The long, club-shaped *Teredina* boring was lined with a calcareous sheath, which accommodated the large intestine needed for the bacteria-aided digestion of the wood filings. T. *suciensis* occurs in the Lambert Formation on Hornby Island and in the Haslam Formation at Brannan Lake. Fossil driftwood is very common in most Cretaceous formations on Vancouver Island and, like recent driftwood in the Strait of Georgia, virtually all of it is riddled by teredo borings.

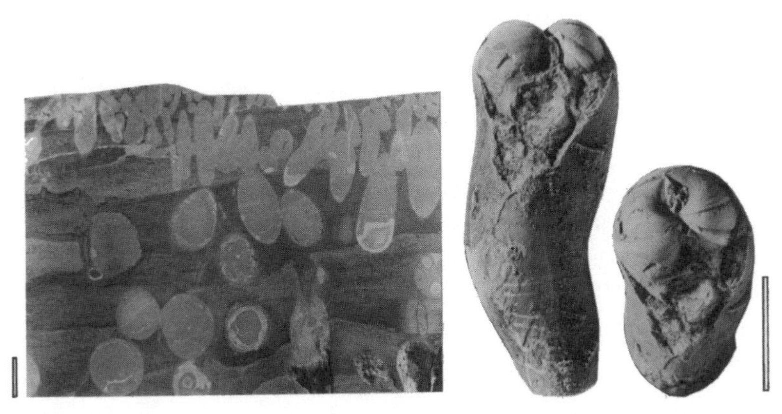

Figure 62 Driftwood (now petrified) bored by the bivalve *Teredina* (VIPM 026) and the bivalve *Teredina suciensis* in its lined boring (VIPM 027). Both from Lambert Fm., Collishaw Point, Hornby Island.

Martesia (Figure 63). *Martesia* is a wood-boring bivalve that differs from *Teredina* in having much larger valves, each with a single conspicuous ridge and fine transverse ridges. Unlike *Teredina*, *Martesia* does not line its borings with calcareous material. *M. clausa* has been found in coalified wood from the Haslam Formation of the Nanaimo area.

Figure 63 The bivalve *Martesia clausa* (VIPM 028) and coalified wood with *Martesia* borings (VIPM 029). Both from Trent River Fm., Northwest Bay.

Anomia (Figure 64). Jingle shells of the genus *Anomia* are quite common in modern tropical and temperate seas, where they live on rocks or affixed to other shells by a strong calcified byssus that extends through a hole in the lower valve. *A. vancouverensis* occurs widely in the Trent River, Haslam, and Lambert formations on Vancouver and Hornby islands. These bivalves frequently take on the sculpture of the shell to which they are attached. For example, the fine ribbing of the ammonite *Neophylloceras* is faint, but clearly visible on the cemented valve of the illustrated specimen of *Anomia*.

Pinna (Figure 65). Cretaceous pen shells are long and slender, with conspicuous, sharp-edged, longitudinal ribs. *Pinna calamitoides* is a rare bivalve in the Trent River Formation at Northwest Bay.

Figure 64 The bivalve *Anomia vancouverensis*, upper and lower valves (VIPM 030, 031) and attached to the ammonite *Neophylloceras* (VIPM 032). All from Haslam Fm., Brannan Lake.

Figure 65 The bivalves *Pinna calamitoides* (VIPM 033, 034), Trent River Fm., Northwest Bay and *Acila shumardi* (VIPM 035, 036), Haslam Fm., Brannan Lake.

Acila (Figure 65). Shales of the Trent River Formation and Haslam Formation contain numerous specimens of this bivalve. When cracked out of the shale, they appear to be featureless shells, but close inspection with a hand lens reveals the striking sculpture of ribs arranged in steep-sided chevrons that cross the faint growth lines. *Acila shumardi* is a small bivalve, generally less than 1 cm long.

Idonearca (Figure 66). Heavy, quadrate, box-like bivalves with thick shells found in the shales of the Haslam and Cedar District formations are assigned to *Idonearca*. The teeth are arranged in a transverse row. The ligamental area above the hinge margin bears prominent chevron-shaped grooves. *I. truncata* has sculpture of closely spaced radiating lines interrupted by sharp growth lines. *Idonearca* must have been a slow burrower in mud.

Figure 66 The bivalve *Idonearca truncata* (RBCM 91813c), Cedar District Fm., Sucia Island and in cross-section (VIPM 037), Haslam Fm., Brannan Lake.

Nemodon (Figure 67). Like mussels, modern arc shells such as *Arca* live fixed to rocks or shells by a strong byssus. The Cretaceous *Nemodon* is virtually identical and undoubtedly it lived in the same way. *N. vancouverensis* is quite common in shales of the Trent River and Lambert formations. This species is one of a handful of fossil mollusks collected in the 1850s near the coal mine that had just opened in Nanaimo. They were described in 1857 by the American paleontologist Fielding

Figure 67 The bivalve *Nemodon vancouverensis* (VIPM 038, 039), Lambert Fm., Collishaw Point, Hornby Island.

B. Meek, who concluded correctly that these fossils and, more importantly at the time, the coal beds are of Cretaceous age and not, as previously thought, of Cenozoic age.

Solemya (Figure 68). *Solemya*, the awning clam, first appeared more than 400 million years ago in the Devonian and is probably the oldest bivalve genus still living. This living fossil is oblong, gaping at both ends, thin-shelled, with subdued growth lines and conspicuous double ridges that radiate out from the umbo. In life, a broad fringe of horny material extends around the margin of the shell—hence the common name. An awning clam lives by circulating sea water through a deep, U-shaped burrow. *Solemya* sp. has only been found in the Lambert Formation on Hornby Island.

SCAPHOPODS

Antalis (Figure 68). The tusk shells are carnivorous mollusks with tapering, curved, calcareous shells that are open at both ends. In life, the wide end was buried in sediment and the animal employed long tentacles to search for small prey species in the sediment. *Antalis* is a simple, smooth scaphopod with fine growth lines. Two species are found in Cretaceous rocks of Vancouver Island: *A. nanaimoense* from the Trent River Formation is more slender than *A. cooperi* from the Lambert Formation.

Figure 68 Top left: the bivalve *Solemya* sp. (VIPM 040), Lambert Fm., Collishaw Point, Hornby Island. Bottom left: the gastropod *Capulus corrugatus* (RBCM) from the Trent River Fm., Puntledge River. Right: the scaphopod *Antalis cooperi* (VIPM 041, 042, 043), Lambert Fm., Collishaw Point, Hornby Island.

GASTROPODS

Capulus (Figure 68). The shell of this elf-cap snail has broad concentric folds on the sides with superimposed fine growth lines. *Capulus corrugatus* is a fairly common fossil in the Trent River Formation.

Bernaya (Figure 69). At the present time, cowries are attractive glossy snails confined to warm seas. In a living cowrie, the latest whorl entirely envelops the earlier formed whorls, giving a long plump shape to the snail. The aperture is a long, narrow, toothed slit. In the Upper Cretaceous, *Bernaya*, one of the first members of the cowrie family Cypraeidae, the first-formed whorls can still be seen and the aperture is smooth. *B. crawfordcatei* is a rare snail that has only been found in the Haslam Formation at Brannan Lake.

Eoacteon (Figure 69). *Eoacteon* is a small, smoothly spired, egg-shaped snail with faint spiral sculpture and an extended aperture. *E.* cf. *linteus* is an uncommon fossil in the Haslam Formation at Brannan Lake.

Figure 69 The gastropods *Bernaya crawfordcatei* (VIPM 044) and *Eoacteon* cf. *linteus* (VIPM 045). Both from Haslam Fm., Brannan Lake.

Hemiacirsa (Figure 70). *Hemiacirsa* is a long, slender, pointed shell with rounded whorls and a sculpture of strong transverse ribs and fine spiral threads. H. *newcombii* was named after Dr. C.F. Newcombe, who collected west coast artifacts and natural history specimens for the Provincial Museum in Victoria, now the Royal British Columbia Museum. This species has been collected from the Haslam Formation near Nanaimo and from the Cedar District Formation on Sucia Island. This high-spired snail is a member of the family Epitoniidae—the wendletraps, whose peculiar name was derived from the German Wendeltreppe, "a spiral staircase."

Pseudocymia (Figures 70 and 71). *Pseudocymia* was one of the newly evolved carnivorous snails that appeared in the Late Cretaceous of the west coast. It has a blunt, spindle-shaped shell sculpted by a spiral row of prominent nodes and by dense spiral threads. At least two species occur on Vancouver Island: *Pseudocymia* sp. from the Haslam Formation of Brannan Lake and P. *cahilli* from the Cedar District Formation of Sucia Island.

Oligoptycha (Figure 71). This unmistakable fossil snail consists of a small rotund shell with a depressed spiral and carrying a sculpture of closely aligned spiral ribs connected by fine, net-like lines. The aperture is outlined by a prominent thick lip, which may carry fine corrugations. *Oligoptycha corrugata* occurs in the Haslam Formation of Brannan Lake.

Figure 70 The gastropods *Hemiacirsa newcombii* (VIPM 046, 047), Trent River Fm., French Creek and *Pseudocymia* sp. (VIPM 048), Haslam Fm., Brannan Lake.

Figure 71 The gastropods *Pseudocymia cahilli* (RBCM 91811c, 91811d), Cedar District Fm., Sucia Island; *Oligoptycha corrugata* (VIPM 049, 050, 051), Haslam Fm., Brannan Lake; and *Nonacteonina* sp. (VIPM 052), Lambert Fm., Collishaw Point, Hornby Island.

Nonacteonina (Figure 71). This slender, awl-like fossil looks more like a miniature narwhal tusk than the snail it is. *Nonacteonina* sp. from the Lambert Formation on Hornby Island bears fine transverse ribs crossed by fine spiral sculpture.

Longoconcha (Figure 72). A new group of carnivorous gastropods, the Volutacea, appeared in the Late Cretaceous and quickly diversified into a large number of genera; *Longoconcha* was one of them. It is a long, spindle-shaped snail with a sharply tapering spire and coarse net-like sculpture. The narrow aperture is extended into a long siphonal canal and the central column has three strong spiral folds (only seen in a cross-section). *L. navarroensis* from the Haslam Formation is a beautiful example of these volute snails.

Figure 72 The gastropod *Longoconcha navarroensis* (VIPM 053, P. Bock Collection, VIPM 054), Haslam Fm., Brannan Lake.

Figure 73 The gastropod *Anchura callosa* (P. Bock Collection, VIPM 055), Trent River Fm., Northwest Bay.

Anchura (Figure 73). The family Aporrhaidae, whose living members are known as pelican's foot shells, are impressive snails with a high, turreted spire and an elaborately expanded aperture. For most of its life, the Upper Cretaceous *Anchura* would have looked like a normal high-spired snail with a net-like sculpture formed of ribs and spiral threads. Then, late in life, after nine or ten whorls, this snail secreted a sharp ridge, which became extended as a lateral flange and then into a posterior spine nearly as long as the turreted spire. At the same time, the shell was extended into a long, slender, siphonal canal. A. *callosa* is a rare snail in the lower Trent River Formation at Northwest Bay.

Tessarolax (Figure 74). It is difficult to imagine a stranger-looking fossil than *Tessarolax*. Despite a superficial similarity to a sea star with five arms, this Upper Cretaceous fossil is a snail. The ill-defined coiled central portion of the shell is surrounded by five long rigid projections that lie in about the same plane. Each projection has a knobby elbow bend. On the lower side, each spiny projection has a central canal extending to the tip and the coiled shell is obscured by a large flat pad. T. *distorta* is a common fossil in shales of the Trent River and Lambert formations, but few specimens are anywhere near complete. Most of the fossils are millimetre-sized broken pieces of the spiny projections.

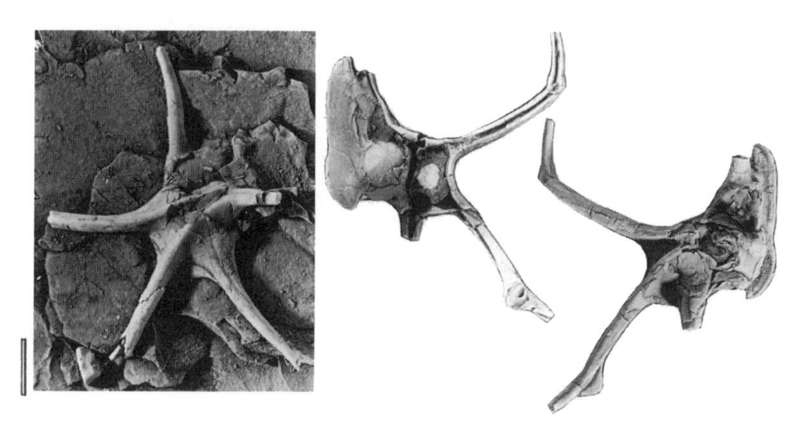

Figure 74 The gastropod *Tessarolax distorta* (VIPM 056, 057), Lambert Fm., Collishaw Point, Hornby Island.

NAUTILOIDS

Eutrephoceras (Figure 75). Nautiloids survived the great extinction that terminated all the ammonites at the end of the Cretaceous. The pearly *Nautilus* from the east Pacific Ocean is the only nautiloid genus alive today. The Cretaceous *Eutrephoceras*—involute and compressed, with gently curving sutures—is so similar to *Nautilus* that it was probably ancestral to the living genus. *E. campbelli*, which reaches a diameter of 25 cm, is quite common in the Trent River Formation and in the Haslam Formation. A few specimens of this nautiloid carry unmistakable evidence of attempted predation by mosasaurs—circular punctures made by peg-like teeth.

AMMONITES

Polyptychoceras (Figure 76). The name of this heteromorph ammonite is Greek for "many-folded horn," and the name is apt. Local collectors call them "candy canes." Coiled like a trombone, this ammonite consists of at least three parallel shafts, the inner one smooth and the

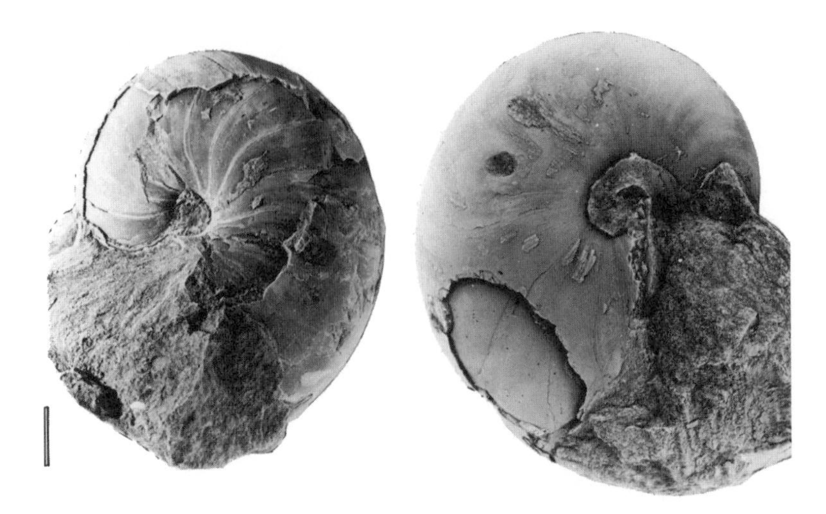

Figure 75 The nautiloid *Eutrephoceras campbelli* (VIPM 058, J. Haegert Collection), Haslam Fm., Brannan Lake.

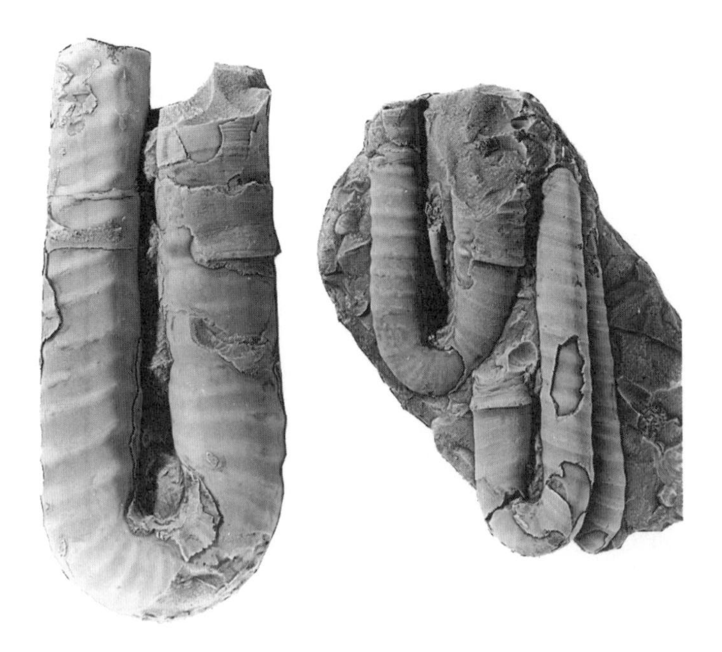

Figure 76 The heteromorph ammonite *Polyptychoceras vancouverense* (VIPM 060, 061), Trent River Fm., Trent River.

outer two with subdued rounded ribs. A single, prominent, ledge-like rib is located just below the rim of the mature living chamber. P. *vancouverense*, the only species, is restricted to the lower part of the Trent River Formation. By far the best specimens come from concretions in shales exposed on the Trent River. The ammonite occurs in two sizes—the shaft of the smaller is half the diameter of the larger. It is possible that the larger shells housed females and the smaller shells males.

Eupachydiscus (Figure 77). This large, robust, coarsely ribbed ammonite is one of the most characteristic fossils in the siltstones of the lower Trent River Formation, and is exposed on the Puntledge and Browns rivers. The shell material of most specimens is soft and chalky, or else missing to reveal the complexly fluted sutures underneath. Good specimens of *Eupachydiscus* have become quite difficult to find over the last few years—frequently all that remains is the ghostly external impression of the ammonite left by a more fortunate collector. E. *perplicatus* bears rounded ribs of the same size. E. *haradai* has smaller, sharp-edged ribs inserted between the rounded ones.

Figure 77 The ammonite *Eupachydiscus perplicatus* (CDM 011), Trent River Fm., Stotan Falls, Puntledge River.

Bostrychoceras (Figure 78). This curious heteromorph ammonite is restricted to the lower Trent River and Haslam formations. The evocative name is Greek for "ringlet horn." B. *elongatum*, the sole Vancouver Island species, consists of an initial spiral of five to seven helical whorls followed by an upwardly recurved adult portion. The sculpture consists of sharp ribs—slanted on the spiral part and cross-wise on the recurved part. This species displays considerable variation—it spirals either to right or left, apparently in equal numbers, and the adult portion may curve up beside the last spiral whorl or it may curve up beneath it. Fragments of *Bostrychoceras* are quite common, but complete specimens of this large fossil are extremely rare. Entire specimens are known only from the Haslam Formation exposed on Elkhorn Creek southwest of Nanaimo. Here, the tough siltstones release their fossil treasures only with the greatest difficulty.

Figure 78 The heteromorph ammonite *Bostrychoceras elongatum* (J. Haegert Collection), Haslam Fm., Elkhorn Creek.

Figure 79 The heteromorph ammonites *Hyphantoceras* sp. (CDM 012) and *Glyptoxoceras subcompressum* (VIPM 062, 063). All from Trent River Fm., Puntledge River.

Hyphantoceras (Figure 79). Superficially similar to *Bostrychoceras*, this small heteromorph ammonite consists of a loosely coiled spiral with a sculpture of flat-topped ribs. It is a rare fossil on Vancouver Island; only a single specimen of *Hyphantoceras* sp. from the lower Trent River Formation is known.

Glyptoxoceras (Figure 79). This delicate, thin-shelled heteromorph is an extremely common ammonite in the shales of the lower Trent River and Haslam formations. Unfortunately, it is almost invariably represented by 1- or 2-cm-long crushed and broken fragments. Complete specimens have not yet been collected on Vancouver Island, but large fragments give a good idea of the shape of the entire shell. *Glyptoxoceras subcompressum* starts as a small, horizontally coiled helix. After three or four corkscrew turns, it begins to coil in broad, open, circular whorls that nearly touch one another. Throughout, the shell is sculpted by dense, sharp-edged ribs.

Hauericeras (Figure 80). This extremely compressed ammonite—"flying saucer" to collectors—is quite common in concretions from the lower Trent River and Haslam formations. The largest specimen we

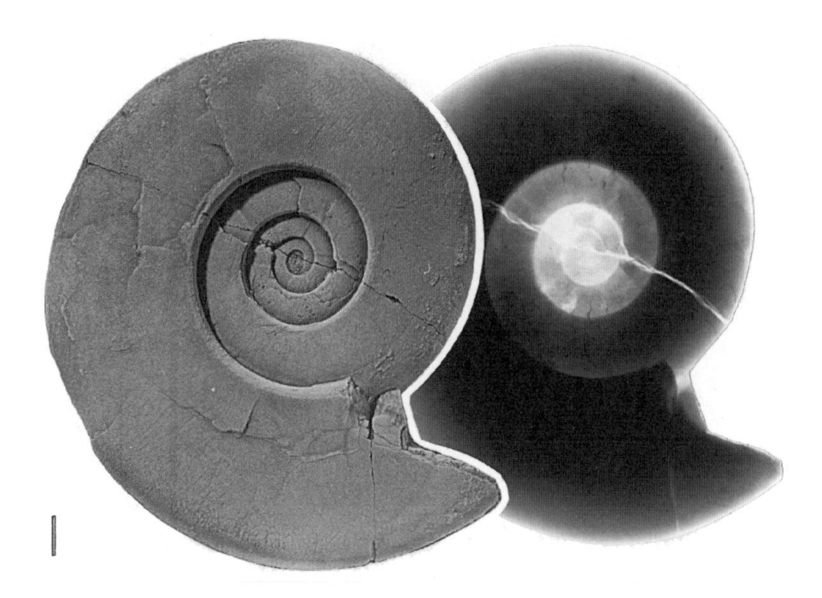

Figure 80 The ammonite *Hauericeras gardeni* (VIPM 064), side view and radiograph, Haslam Fm., Brannan Lake.

have collected has a diameter of 25 cm, but it is only 3 cm deep. Widely spaced constrictions are evident on the juvenile shell, but these disappear on the mature portion, which is smooth with an acutely angled keel. *Hauericeras gardeni* is the only species of this ammonite genus on Vancouver Island.

Cyphoceras (Figure 81). Looking more like a horn about to be tootled by a Dr. Seuss character than a Cretaceous fossil, this heteromorph ammonite consists of up to six parallel shafts joined by U-shaped turns. Earlier-formed shafts carry simple transverse ribs which, in later shafts, are augmented by widely spaced marginal spines. *C. lineatum* is a rare ammonite in the Trent River Formation. It is large, up to 65 cm across and, because it is delicate and thin shelled, only a single moderately complete specimen has ever been collected.

Ryugasella (Figure 82). Pencil-stub-sized fragments of this straight heteromorph are quite common in the lower Trent River Formation, but because complete specimens are not known, the shape of this

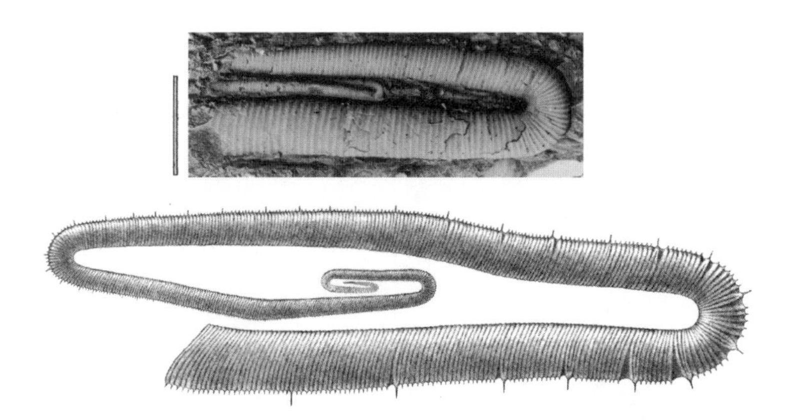

Figure 81 The heteromorph ammonite *Cyphoceras lineatum*, inner four shafts (J. Haegert Collection), Trent River Fm., Northwest Bay. Reconstruction of entire *C. lineatum*, 65 cm across (modified from Ward and Mallory, 1977).

Figure 82 The heteromorph ammonite *Ryugasella ryugasensis* (CDM 013, VIPM 067), Trent River Fm., Puntledge River and Trent River.

peculiar ammonite remains a mystery. *Ryugasella ryugasensis* bears dense transverse ribs interrupted by the occasional constriction.

Gaudryceras (Figure 83). *Gaudryceras* is an evolute ammonite with a fine sinuous sculpture of ribs, or lirae. Two species are known in this area. G. *striatum* from the Haslam and lower Trent River formations bears widely spaced rounded ribs in addition to fine, dense lirae. G. *denmanense* from the Lambert Formation on Denman and Hornby islands has closely spaced, sharp ribs, but no lirae. *Gaudryceras* is notable in that the entire growth history of the ammonite can be seen clearly on a single specimen, all the way back to the protoconch, the bulbous, first-formed part of the shell. *Gaudryceras* is rare in Upper Cretaceous rocks on Vancouver Island, but it has a worldwide distribution.

Figure 83 The ammonite *Gaudryceras striatum* (CDM), Haslam Fm., Brannan Lake.

Pseudoschloenbachia (Figures 84 and 85). This ammonite has probably the most juicily satisfying name to pronounce of all the west coast fossils. *Pseudoschloenbachia* is a compressed ammonite that bears a characteristic array of nodes flanking a sharp, serrated keel and on the sides. The low, sweeping ribs become progressively subdued on the mature parts of the shell. *Pseudoschloenbachia* is represented by a single species on Vancouver Island. *P. umbulazi* has been found only in concretions in the Haslam Formation at Brannan Lake.

Damesites (Figure 85). This smooth involute ammonite with regular, wavy constrictions is very similar to *Desmophyllites*. The low, conspicuous keel, however, easily distinguishes *Damesites*. *D. sugata*, the only species, is uncommon in the Haslam and lower Trent River formations. It almost always occurs in shales, rarely in concretions.

Figure 84 The ammonite *Pseudoschloenbachia umbulazi* (VIPM 069), Haslam Fm., Brannan Lake.

Figure 85 The ammonites *Damesites sugata* (VIPM 072), Trent River Fm., Puntledge River and *Pseudoschloenbachia umbulazi* (VIPM 073), Haslam Fm., Brannan Lake.

Epigoniceras (Figure 86). This ammonite bears a curious similarity to a badly worn radial tire. E. *epigonum* is a smooth, rather evolute ammonite from the Haslam and lower Trent River formations, where it occurs only in small concretions. It is readily distinguished by the characteristic square outline of its living chamber and by the three low ridges that run along the venter.

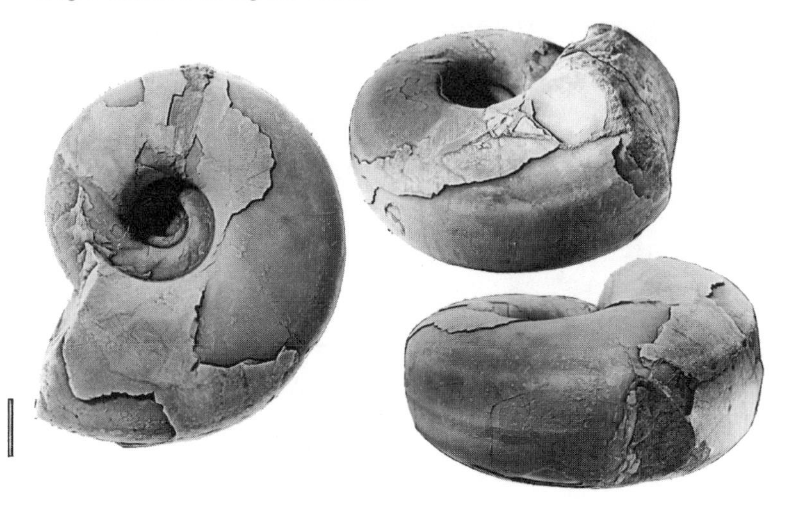

Figure 86 The ammonite *Epigoniceras epigonum* (VIPM 070, 071), Haslam Fm., Brannan Lake.

Canadoceras (Figures 87, 88, and 89). This "Canadian horn" is closely related to E*upachydiscus*, from which it differs by being more compressed and by having rounded ribs that are closely spaced. C. *yokoyamai* is moderately common at many localities of the Haslam Formation; it is particularly common at Brannan Lake outside Nanaimo, where specimens as large as 50 cm in diameter have been collected. This ammonite is easy to spot in the field because it appears bright white—the original shell material has been altered to a white chalky substance. C. *newberryanum* (Figure 89), which occurs in the Cedar District Formation and in the Oyster Bay Formation, has regular, trough-like constrictions and displays a tendency to uncoil at later stages. Unusual U-shaped folded elements with wrinkled sides occur in concretions at Brannan Lake. Elsewhere, similar fossils are

Figure 87 The ammonite *Canadoceras yokoyamai* (VIPM 074), Haslam Fm., Brannan Lake.

Figure 88 Upper and lower jaws of *Octopus dofleini* (VIPM 076), caught off the dock at Campbell River and lower (?) jaws of the ammonite *Canadoceras yokoyamai* (VIPM 077, 078), Haslam Fm., Brannan Lake.

Figure 89 The ammonite *Canadoceras newberryanum* (J. Haegert Collection), Cedar District Fm., Sucia Island.

known as aptychi, which now are recognized to be the lower jaws of ammonites. In all likelihood, these belong to *Canadoceras*. In Figure 88, the lower jaw of *C. yokoyamai* is shown alongside the jaws of a living cephalopod—a species of octopus from the Strait of Georgia. It is not known whether this ammonite had only a lower jaw or identical upper and lower jaws.

Neophylloceras (Figure 90). Phylloceratids are an unusually conservative group of ammonites that persisted from the Triassic to the Cretaceous with only minor change, a time period of about 170 million years. *Neophylloceras* is one of the last members of this long-lived family. It is an involute compressed ammonite with a rounded keel and a sculpture of fine, closely spaced lirae. The shells, preserved with iridescent nacre, are most commonly found in concretions. Two species occur in Upper Cretaceous rocks of Vancouver Island: *N. ramosum* from the Trent River, Haslam, and Lambert formations has flat flanks; and *N. surya* from the Lambert Formation has conspicuously undulating flanks.

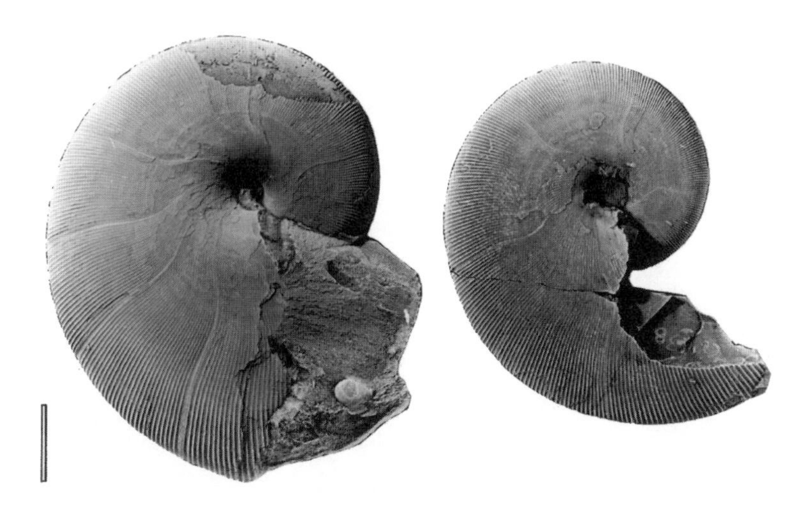

Figure 90 The ammonite *Neophylloceras ramosum* (VIPM 079), Haslam Fm., Brannan Lake and (VIPM 080), Lambert Fm., Collishaw Point, Hornby Island.

Nostoceras (Figures 91 and 92). Because of their bizarre shapes, Cretaceous heteromorph ammonites from the Gulf Islands and Vancouver Island have been the specific target of fossil collectors. *Nostoceras* from Hornby Island is probably the heteromorph most eagerly sought by both amateur and professional paleontologists. This ammonite consists of an expanding, spirally coiled turret that is followed by a downwardly hanging living chamber that recurves upward; hence the name *Nostoceras*, Greek for "the horn that returns home." The sculpture consists of sharp-edged ribs, many of which split and recombine. *Nostoceras* is fairly common in concretions at Collishaw Point, but complete specimens are almost unknown. That illustrated in Figure 91 is the only complete specimen known on the west coast. Other specimens consist of either the turret or the U-shaped living chamber. At least three different species occur in the Lambert Formation on Hornby Island: *N. hornbyense* has a low, dextrally coiled turret; *N. perversum* (Figure 92) is sinistrally coiled, with denser ribs; *Nostoceras* sp. (Figure 92) has a high, dextrally coiled turret. *N. hornbyense* may be yet another of the ammonites displaying sexual dimorphism—the microconch being less than half the size of the macroconch.

Figure 91 The heteromorph ammonite *Nostoceras hornbyense* (CDM), Lambert Fm., Collishaw Point, Hornby Island.

Figure 92 The heteromorph ammonites *Nostoceras perversum* (VIPM 082) and *Nostoceras* sp. (VIPM 083, 084). All from Lambert Fm., Collishaw Point, Hornby Island.

Stantonoceras (Figure 93). This thorny ammonite is extremely rare in the west coast Cretaceous—only a single specimen is known from the Cedar District Formation on Sucia Island. S. cf. *guadeloupae* has an almost flat keel and sides that are studded with short spines connected by broad ribs. *Stantonoceras* is, essentially, a spiny *Hoploplacenticeras* (see Figure 99).

Figure 93 The ammonite *Stantonoceras* cf. *guadeloupae* (J. Haegert Collection), Cedar District Fm., Sucia Island.

Anagaudryceras (Figure 94). The inner whorls of *Anagaudryceras* comprise a dense spiral swirling away from the globular protoconch, the first-formed part of the ammonite. A single species is known: A. *politissimum* from concretions in the Lambert Formation of Hornby Island. This ammonite lacks fine sculpture, but has widely spaced constrictions that run obliquely across the shell. It is extremely difficult to extract specimens of *Anagaudryceras* from concretions without breaking them.

Desmophyllites (Figure 95). Most specimens of this ammonite with a rounded profile come from the Cedar District Formation of Sucia

Island. D. *diphylloides* (Forbes), the only species, is a large, smooth, involute ammonite with regularly spaced S-shaped constrictions that appear more prominent on the interior than on the exterior of the shell. This ammonite is superficially similar to *Damesites* (Figure 85), but *Desmophyllites* lacks a keel.

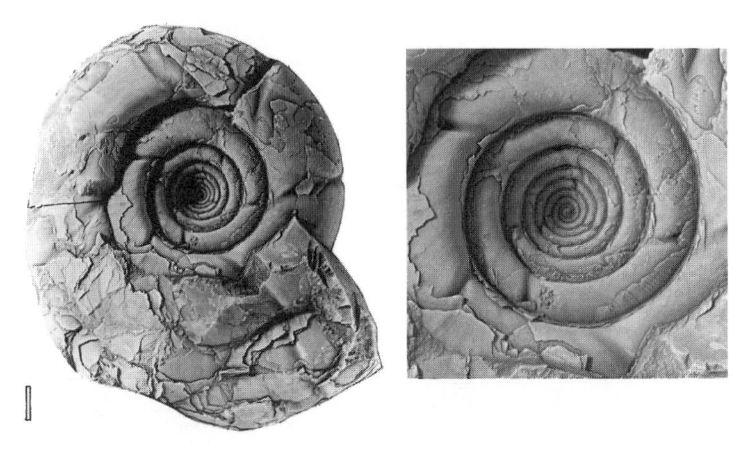

Figure 94 The ammonite *Anagaudryceras politissimum* (VIPM 085), Lambert Fm., Collishaw Point, Hornby Island.

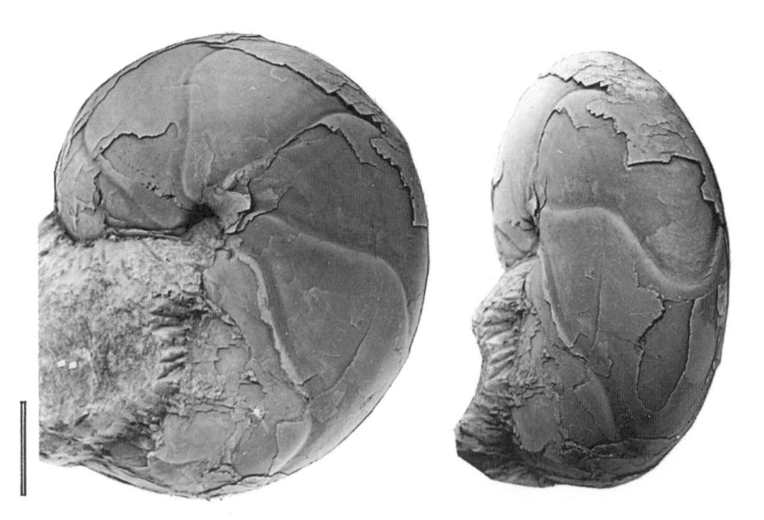

Figure 95 The ammonite *Desmophyllites diphylloides* (J. Haegert Collection), Cedar District Fm., Sucia Island.

Figure 96 The heteromorph ammonite *Diplomoceras cylindraceum* (VIPM 086, 087), Lambert Fm., Collishaw Point, Hornby Island.

Diplomoceras (Figure 96). Straight segments of D. *cylindraceum* are characteristic fossils in the Lambert Formation on Hornby Island—large specimens looking like half-metre-long pieces of ribbed cordwood. This heteromorph ammonite cannot be reconstructed with full confidence because the most complete specimens known are 40-cm-long U- or V-shaped portions of the shell. In life, it was probably at least a metre across and must have looked something like a finely corrugated tuba. The earliest-formed shell has not yet been found. The smaller shafts are covered by a sculpture of fine transverse ribs interrupted by regular constrictions. Only the fine ribbing occurs on the adult shell. Even small fragments of this ammonite are attractive because they frequently preserve the original mother-of-pearl nacre.

Figure 97 The heteromorph ammonite *Baculites pacificus* (VIPM 088, 089), Oyster Bay Fm., Shelter Point and a juvenile specimen of *Baculites*, 10 mm long , showing initial coiling (K. Morrison Collection), Lambert Fm., Collishaw Point, Hornby Island.

Baculites (Figure 97). Fragments of *Baculites* are probably the most common ammonite in Upper Cretaceous rocks of Vancouver Island and the Gulf Islands. This heteromorph ammonite starts as a few minute spiral whorls before straightening into a gradually expanding linear shaft that bears low, oblique or S-shaped ribs and fine growth lines. The initial whorls are almost never found. Complete specimens can reach a length of a metre or more. The best specimens are found inside long, sausage-shaped concretions at Shelter Point south of Campbell River and at Collishaw Point on Hornby Island. Some half-dozen species of *Baculites* have been recognized in the west coast Cretaceous. Of these, B. *bailyi*, B. *chicoensis*, B. *inornatus*, and B. *rex* are smooth, with very fine growth lines, and have egg-shaped cross-sections. B. *pacificus* and B. *occidentalis* bear broad oblique ribs in addition to the growth lines, and are oval in cross-section. *Baculites* is a colourful iridescent ammonite. Many specimens, particularly those from the Lambert Formation on Hornby Island, retain the original mother-of-pearl shell.

Pseudophyllites (Figure 98). A smooth involute ammonite with a rounded and rapidly expanding living chamber, *Pseudophyllites indra* is a rare species in the Lambert Formation on Hornby Island. The shell material is generally brown. Where it is well preserved, one can see that it is not really smooth at all, but instead covered by a network of fine ridges. P. *indra* has a worldwide distribution in the Late Cretaceous.

Submortoniceras (Figure 98). The coarse wavy ribs on the flanks of this rare compressed involute ammonite are diagnostic, as are the two rows of dash-shaped tubercles flanking a low central ridge on the keel. *Submortoniceras* sp. occurs in the Oyster Bay Formation at Shelter Point.

Figure 98 The ammonites *Pseudophyllites indra* (VIPM 090), Lambert Fm., Collishaw Point, Hornby Island and *Submortoniceras* sp. (VIPM 091). Oyster Bay Fm., Shelter Point.

Hoploplacenticeras (Figure 99). Even fragmentary specimens of this compressed involute ammonite are uncommon. H. *vancouverense*, the only species known on the west coast, is found only in the Cedar District Formation of Sucia Island. It bears low radial ribs and paired elongate spines flanking the flat keel. The photographed specimen

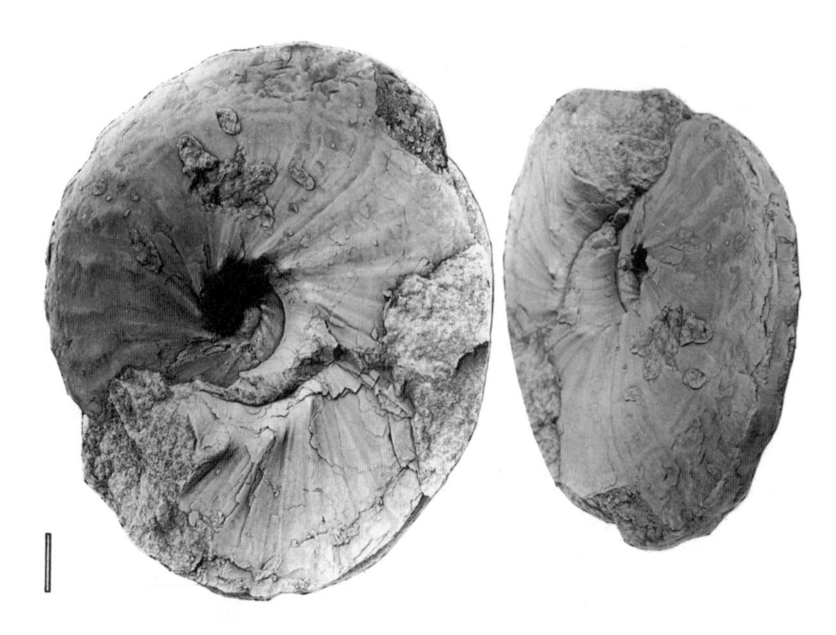

Figure 99 The ammonite *Hoploplacenticeras vancouverense* (J. Haegert Collection), Cedar District Fm., Sucia Island.

shows characteristic shell damage, virtually identical to that seen on other compressed ammonites (for example, *Pseudoschloenbachia*, Figure 84, and *Stantonoceras*, Figure 93). In all of these ammonites, the thin shell of the back part of the living chamber has been sliced open. Although it cannot be determined with certainty, the damage to these ammonites probably occurred post-mortem and was done by scavenging crabs.

Pachydiscus (Figures 100 and 101). Many circular concretions in the Lambert Formation at Collishaw Point on Hornby Island are formed around large specimens of *Pachydiscus*. The preservation of this ammonite is generally good, but it is unusual in that, in virtually every case, only the lower half of the ammonite is preserved. Presumably, because it was not buried in mud, the upper half of the shell was dissolved while it was lying on the sea floor. *Pachydiscus* is the only Cretaceous ammonite from Vancouver Island that bears evidence of reptile predation—centimetre-sized mosasaur teeth marks (see Figure

Figure 100 The ammonite *Pachydiscus suciaensis* (VIPM 151, 152), Lambert Fm., Collishaw Point, Hornby Island.

Figure 101 The ammonite *Pachydiscus hornbyense* (VIPM 092), Lambert Fm., Collishaw Point, Hornby Island.

129). This ammonite must have attained impressive sizes—P. *suciaensis* frequently reaches diameters of 30 cm or more. P. *hornbyense* is a smaller ammonite with sparse asymmetric ribs across the venter.

SEPIIDS

New cuttlefish (Figure 102). Living cuttlefish of the genus *Sepia* have ten arms, two much longer than the rest, all equipped with suckers. These active marine carnivores have a flattened body and well-developed lateral fins. They maintain neutral buoyancy by controlling

Figure 102 The cuttlebone of a new, unnamed cuttlefish (B. Geppert Collection), Lambert Fm., Collishaw Point, Hornby Island and the cuttlebone of the living cuttlefish *Sepia officialis* bought at a pet-food store (VIPM 059).

gas pressure in the chambers of their internal shell. This shell is familiar to bird fanciers as the porous cuttlebone sold in pet supply stores. The oldest cuttlebones known in the fossil record have been collected from the Lambert Formation at Collishaw Point on Hornby Island. This fossil cuttlefish has no formal name as yet, but it is clearly similar to *Sepia*, although much larger and carrying a dozen strong longitudinal ribs and a scalloped front margin.

Naefia (Figure 103). A tiny perfect unrolled cornucopia, *Naefia* is actually the internal chambered shell of a small Late Cretaceous sepiid. In life, it must have looked rather like the living cuttlefish *Spirula*, the so-called ram's horn shell, which has a similar internal shell, but one in the form of an open coil. *Naefia* n.sp. is an extremely rare fossil in concretions from the Lambert Formation at Collishaw Point on Hornby Island, where it is preserved as part of the fossilized feces of bony fishes.

Figure 103 The internal shells of the cuttlefish *Naefia* n. sp. in a fish coprolite (both sides of the split concretion are shown) (VIPM 153). Lambert Fm., Collishaw Point, Hornby Island.

INSECTS

Weevil (Figure 104). Perhaps apocryphal, but when asked by a theologian what his extensive knowledge of the natural world has taught him about the mind of the Creator, the great British biologist J.B.S. Haldane is supposed to have responded: "[He has] an inordinate fondness for beetles." There are certainly a lot of them—with 400,000 named species, more Coleoptera are known than any other animal group. Fossil beetles are fairly common in Cenozoic lake deposits and in amber, but, not surprisingly, they are very rare in marine rock, particularly those of Cretaceous age. A thin shale unit in the Lambert Formation at Collishaw Point on Hornby Island yields the only fossil beetle known from a Cretaceous rock of the west coast. This fossil consists of the wing covers (elytra) of a member of the family Curculionidae. Each elytron is scored by fine longitudinal furrows that contain minute pits. This family, the weevils, extends back to the Late Jurassic. Weevils are those beetles with the head prolonged into a snout with biting jaws at the end and are known to be very destructive to grain, fruit, cotton bolls, and nuts. Presumably, their habits were similar in the Cretaceous.

Termites (Figure 104). It is not surprising that insect fossils are almost entirely missing from the Cretaceous record on Vancouver Island and

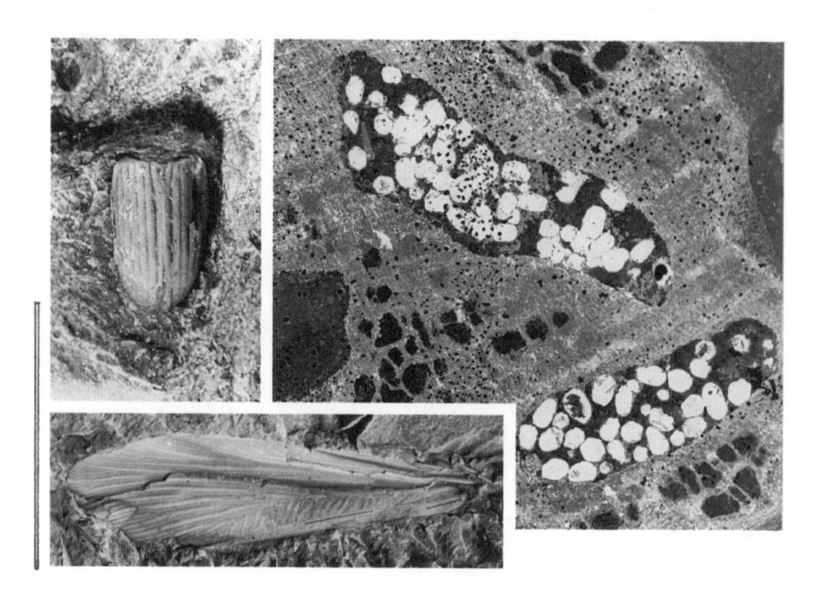

Figure 104 Cretaceous insects. The wing covers of a weevil (K. Morrison Collection), Lambert Fm., Collishaw Point, Hornby Island. Termite galleries excavated in sound conifer wood and packed with termite fecal pellets (here shown white) (VIPM 147), Lambert Fm., Collishaw Point, Hornby Island. Cockroach hind wing in fish coprolite (VIPM 148), Haslam Fm., Brannan Lake.

the Gulf Islands. Insects are, after all, closely tied to terrestrial environments and virtually all of the fossil-bearing Cretaceous rocks are marine. A large piece of fossil driftwood from the Lambert Formation on Hornby Island has provided the first evidence of Cretaceous insect activity in this area. Fossil wood is very common on Hornby and most of it is bored by teredos. But this particular piece of wood is densely riddled with galleries, most packed with millimetre-sized black fecal pellets (which appear white on the photograph). From their size and shape, these galleries are clearly the work of social insects. The size of the excrement pellets, many of which are six-sided, is strongly suggestive of the order Isoptera—termites. If that identification is correct, then the wood must have been worked by termites on the Cretaceous forest floor, swept into a river and then out to sea, where it eventually became waterlogged and sank to the bottom. As the photograph shows, the galleries were excavated in sound, dry wood.

This remarkable fossil from Hornby Island is only the second record of a termite nest preserved in rocks as old as Late Cretaceous. The other one is from Texas. It is evidence of the oldest social activity yet discovered and it provides circumstantial evidence for the presence of intestinal protoctistans, which are necessary to convert cellulose to sugars in the guts of these wood-eating insects.

Cockroach (Figure 104). Cockroaches are genuine living fossils. They were scurrying through the litter on the forest floor for 300 million years before they discovered human habitations. Given the habits and the poor flying ability of these insects, it is not surprising that fossil cockroaches are virtually unknown in marine rocks. And yet, among the broken bits of fish bones and fish spines that comprise a coprolite in a concretion collected from marine shales of the Haslam Formation at Brannan Lake, lay a well-preserved and nearly complete hind wing of a representative of the order Blattodea—roaches. Cretaceous cockroaches are poorly known, so this fossil cannot be identified to genus and species yet. The well-defined veins in the wing will permit identification once the fossil cockroaches of England and Brazil are studied.

We can only speculate on the events that resulted in this insect ending up in the gut of a Cretaceous marine fish. Perhaps the cockroach was blown out to sea and gulped down whole by a fish.

CRABS

Longusorbis (Figure 105). Shales of the Oyster Bay Formation exposed at low tide on Shelter Point 10 km south of Campbell River include numerous sandy calcareous concretions that contain two types of fossils and little else—the straight ammonite *Baculites* and the crab *Longusorbis*. In the Late Cretaceous *Baculites* was cosmopolitan in its distribution. *Longusorbis*, on the other hand, has been collected from only one locality in the world, Shelter Point. But, as if to compensate for its restricted distribution, it occurs there in great numbers—about 800 specimens are presently catalogued at the Vancouver Island Paleontological Museum. *Longusorbis cuniculosus* bears long eye stalks (hence the generic name) and has a distinctly furrowed carapace with three pairs of spines on the front margin and a downwardly turned median tongue. The claws are massive; the right claw is noticeably larger than the left.

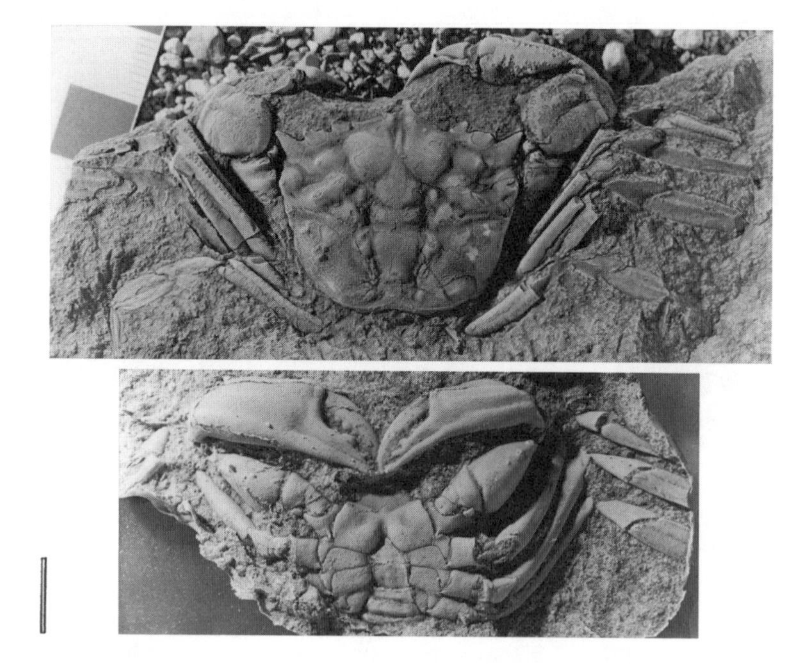

Figure 105 The crab *Longusorbis cuniculosus* (CDM 014, VIPM 094), Oyster Bay Fm., Shelter Point.

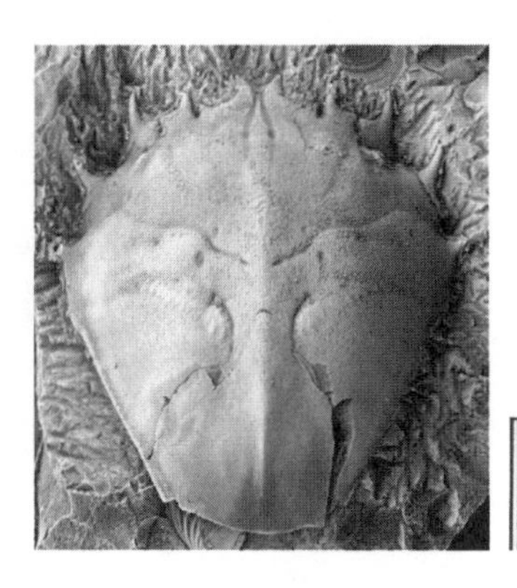

Figure 106 The crabs *Archaeopus vancouverensis* (VIPM 095), Lambert Fm., Collishaw Point, Hornby Island and *Cretacoranina harveyi* (VIPM 096), Trent River, Fm., Trent River.

Archaeopus (Figure 106). Ammonites, bivalves, and wood abound in concretions of the Lambert Formation on Hornby Island to the virtual exclusion of other fossils. A small, distinctly marked fossil crab is fairly common in these beds. *Archaeopus* has an oval carapace with three raised transverse ridges bearing fine granules. The legs and claws are slender and not well calcified. *A. vancouverensis* is the only species occurring in this area.

Cretacoranina (Figure 106). Raninids are atypical crabs because the abdomen, which is permanently bent under the carapace in normal crabs, is free and flexible. At the present time, the family Raninidae is mainly found in the eastern Pacific and Indian oceans where these crabs are burrowers in moderately deep water. *Cretacoranina* occurs in the lower Trent River Formation exposed on the Puntledge and Trent rivers. It has a shield-shaped carapace, without prominent markings, that bears five pairs of sharp spines on its front margin. Flanking a median crease are paired attachment areas of muscles supporting the stomach. *Cretacoranina harveyi* was named for Walter Harvey who, under contract to the Geological Survey of Canada and the Provincial Museum in Victoria, made extensive fossil collections from Vancouver, Denman, and Hornby islands during the 1890s.

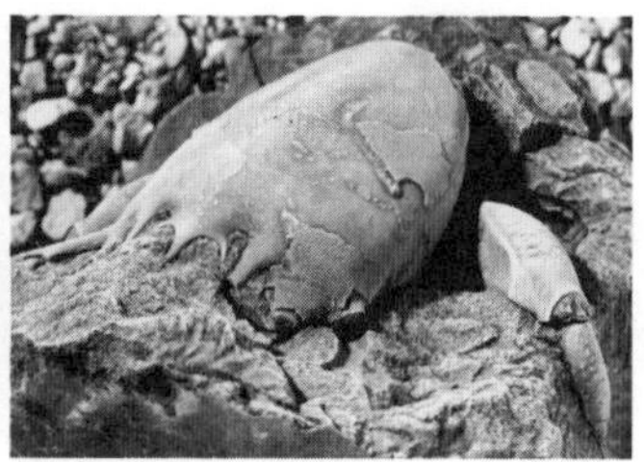

Figure 107 The crab *Rogueus* sp. (VIPM 097), Trent River Fm., Northwest Bay.

Rogueus (Figure 107). Concretions in sandstones of the Trent River Formation at Northwest Bay have yielded a second raninid crab. *Rogueus* sp. has an almost featureless, shield-shaped carapace with three pairs of sharp spines flanking a trident-like central spine. The lateral spines are forked and have the appearance of deer antlers. This crab, like all raninid crabs, bears puny claws. The species is new and yet to be named.

Homolopsis (Figure 108). The tiny square carapace of this rare crab from the Lambert Formation on Hornby Island is reminiscent of an amulet carved to resemble a scarab beetle. Two transverse furrows divide the carapace into three belts, which, in turn, are separated into elliptical and triangular elements. Claws or legs of *Homolopsis* sp. are not known.

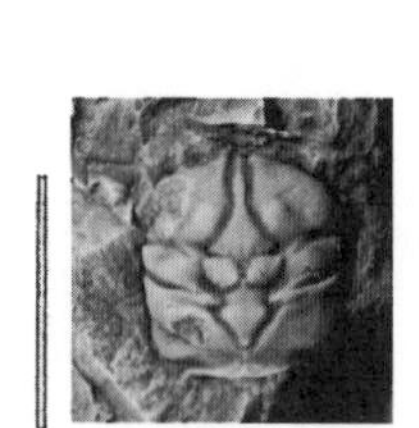

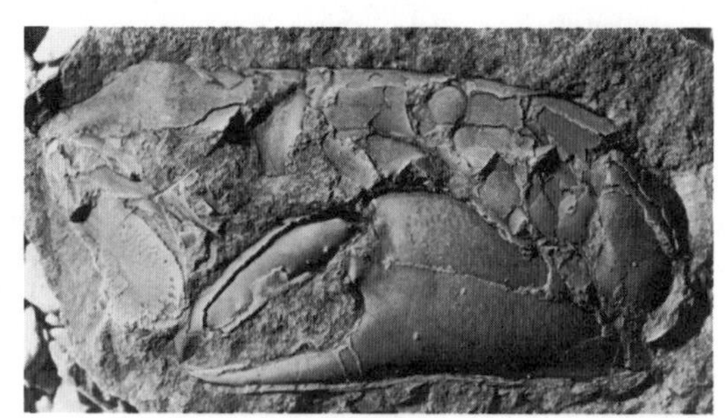

Figure 108 The crab *Homolopsis* sp. (VIPM 098), Lambert Fm., Collishaw Point, Hornby Island and the ghost shrimp *Callianassa whiteavesi* (CDM 015), Trent River Fm., Puntledge River.

GHOST SHRIMP

Callianassa (Figure 108). Just below Stotan Falls west of Courtenay, a flat expanse of siltstones in the middle of the Puntledge River is known as "Callianassa Island" to local collectors due to the abundance of small concretions containing the claws of this crustacean. Because it was mineralized during life, the claw is generally the only part preserved; the carapace and abdomen were delicate and soft, hence the common name, ghost shrimp. The claw is broadly rectangular with delicate pinchers.

Callianassa whiteavesi is one of the few fossils from Vancouver Island named after the paleontologist who described and named the majority of Cretaceous fossils from the west coast. Joseph Frederick Whiteaves was paleontologist to the Geological Survey of Canada for over thirty years. As portrayed in a formal photograph taken late in his life, he was a portly Victorian gentleman with white muttonchops. Although his two monographs on Cretaceous fossils of Vancouver Island and the Gulf Islands were published about 100 years ago, both are still invaluable references. The early GSC directors expected that Whiteaves and his predecessor Elkanah Billings should be knowledgeable about all Cambrian to Cenozoic fossils brought back to Ottawa by numerous mapping parties working from Nova Scotia to British Columbia. With Whiteaves's death in 1909, the fifty-year era of self-taught or apprentice-trained generalist paleontologists at the GSC came to an end. Soon the GSC began to hire university-trained paleontologists who were allowed, indeed expected, to specialize on specific fossils from a small portion of the geologic column.

LOBSTERS

Hoploparia (Figure 109). To most people, a lobster is synonymous with *Homarus americanus*, the Atlantic lobster, but this is only one of perhaps forty-five described genera of lobsters—most known only as Mesozoic or Cenozoic fossils. *Hoploparia* differs only slightly from *Homarus*. It is represented by extraordinarily well-preserved material from the lower Trent River Formation exposed on the Puntledge River, and it also occurs in the Haslam Formation. The cylindrical carapace of this lobster is finely granular with only a few shallow furrows. The

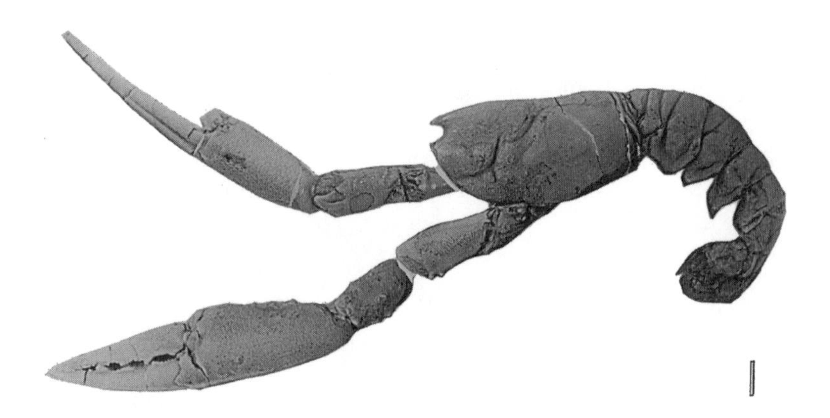

Figure 109 The lobster *Hoploparia* sp. (CDM 001), Trent River Fm., Stotan Falls, Puntledge River.

Figure 110 The lobster *Linuparus vancouverensis* (VIPM 099, 155), Trent River Fm., Stotan Falls, Puntledge River.

long arms end in prominent claws—the crusher claw on the left bears stout teeth and is larger than the nipper claw on the right. The square termination of the second abdominal segment demonstrates that this specimen of *Hoploparia* sp. was a male. A female has a pointed termination. These specimens appear to represent a new, undescribed species. Lobsters have been around for nearly 250 million years. They deserve better treatment than being dropped alive into boiling water!

Linuparus (Figure 110). *Linuparus*, the only spiny lobster found in Cretaceous rocks on Vancouver Island, has a rectangular carapace with three prominent longitudinal ridges on the upper surface. *L. vancouverensis* is moderately common in the lower Trent River Formation on the Browns and Puntledge rivers and in the Lambert Formation on Hornby Island, but complete, well-preserved specimens are rare. Spiny lobsters lack the enlarged front claws of normal lobsters, such as *Homarus* and *Hoploparia*; instead, all five pairs of legs are about the same size. *Linuparus* is a long-lived genus that started in the Cretaceous. This spiny lobster still prowls the sea bottom of the Pacific and Indian oceans.

HEXACORALS

Platycyathus (Figure 111). Rich associations of fossil corals occur in Cretaceous limestones of the southern US, the Mediterranean area, and the Middle East that were deposited in tropical waters. But corals are essentially absent from shales and sandstones of the same age deposited in temperate settings on Vancouver Island. The delicate hexacorals that grew on the large bivalve *Inoceramus* from the Trent River Formation on Englishman River are a rare exception. As shown in the photograph, *Platycyathus* sp. is a low, small, solitary coral with thin septae and whose six-fold symmetry is faintly discernible.

BRACHIOPODS

Cyclothyris (Figure 112). The phylum Brachiopoda was decimated at the great mass extinction that closed the Paleozoic Era. Only a handful of genera of a few families survived into the Mesozoic. The brachiopod *Cyclothyris suciensis* is an uncommon fossil in the Haslam and

Cedar District formations. It bears a prominent beak on the ventral valve and fine sharp radiating ribs.

Figure 111 The solitary hexacoral *Platycyathus* sp. growing on a large *Inoceramus* valve (VIPM 100), Trent River Fm., Englishman River.

Figure 112 The brachiopod *Cyclothyris suciensis* (VIPM 101, 102), Haslam Fm., Brannan Lake.

HEART URCHINS

Hemiaster (Figure 113). Heart urchins display the five-fold radial symmetry typical of their ancestors, the sea urchins, but also have superimposed bilateral symmetry to the petal-like pattern on the upper surface. More-or-less heart-shaped, these echinoderms occupied (and still occupy) deep burrows in the mud where they use their spiny covering to sort through organic-rich sediment and to burrow. *Hemiaster* sp. is known from only a few specimens in the Trent River Formation at Northwest Bay. It has the shape of a flat bun. The length of the petals decreases from the front to the rear.

Figure 113 The heart urchin *Hemiaster* sp. (VIPM), Trent River Fm., Northwest Bay and a sea star belonging to the family Goniasteridae (J. Whittles Collection), Haslam Fm., Maple Bay.

SEA STARS

Goniasteridae (Figure 113). In contrast to their great abundance and diversity in the present seas surrounding Vancouver Island, fossil sea stars are rarely preserved in the rocks of the Island. Only two Cretaceous sea stars are known—a single incomplete arm from the Lambert Formation on Hornby Island and the impression of a relatively complete specimen from the Haslam Formation at Maple Bay south of Crofton. Both belong to undetermined genera of the family Goniasteridae. This family is characterized by large, rectangular marginal plates bordering an interior mosaic of small plates. During life, these plates would have been covered by a tough spiny skin. The short arms of goniasterids tend to be triangular in outline.

SEA URCHINS

Gomphechinus (Figure 115). Sea urchins, or regular echinoids, are common in the Strait of Georgia, where they graze on seaweed in shallow water (and are now the object of a lucrative commercial fishery). Up until now, they have not been found as fossils on Vancouver Island. A single well-preserved individual from the Trent River Formation at Northwest Bay displays the key features of echinoids: the presence of ten radial segments, each consisting of a double row of calcite plates—five pore-bearing segments alternating with five non-perforate segments. Each calcite plate is surmounted by a pair of nipple-like spine bases. This wheel-shaped specimen is a new species of *Gomphechinus*, a genus previously known only from the Cretaceous of Africa.

BRITTLE STARS

Ophiura (Figure 114). "Writhing" is not a word usually applied to fossils, but it seems an appropriate descriptor for the hundreds of fossil brittle stars with whip-like arms seemingly frozen in mid-flail that crowd the bottom of a thin sandstone bed from the Haslam Formation

Figure 114 The brittle star *Ophiura* sp. (J. Whittles Collection), Haslam Fm., Maple Bay.

at Maple Bay, south of Crofton. These brittle stars, or ophiuroids, were smothered and killed by a submarine flow of sediment. Fossil brittle stars are difficult to identify, but these specimens probably belong to *Ophiura*, a living genus with a fossil record extending back to the Cretaceous. The upper side of each arm is covered by three rows of minute plates. These specimens are the first fossil ophiuroids found on Vancouver Island.

CRINOIDS

Marsupites (Figure 115). In 1821 when the British physician J.S. Miller named *Marsupites*, he thought that this bag-like echinoderm was a transitional form linking stalked crinoids and starfish (the name comes from the Greek marsipion, a purse). Even though it lacks a stem and has the appearance of a slightly deflated soccer ball, it is a true crinoid, but a highly unusual one. *Marsupites* is a large globular stemless crinoid composed of 16 polygonal plates, each of which is ornamented by characteristic arrays of lines. Doubtlessly because of their floating habit, the stemless crinoids *Uintacrinus* and *Marsupites* achieved world-wide distribution for the same brief period in the Upper Cretaceous before going extinct. On Vancouver Island, M. *testu-*

Figure 115 The crinoid *Marsupites testudinarius* (C. Ruttan Collection), Haslam Fm., Haslam Creek and the sea urchin *Gomphechinus* n. sp. (VIPM 162), Trent River Fm., Northwest Bay.

Figure 116 The crinoid *Uintacrinus socialis* (CDM 016), Trent River Fm., Stotan Falls, Puntledge River and a reconstruction of the living crinoid. Only the basal parts of the long arms are shown.

dinarius is known from a single locality in the Haslam Formation at Haslam Creek.

Uintacrinus (Figure 116). Dense clusters of large, thin, hexagonal plates in the Trent River Formation constitutes an unusual fossil record of Cretaceous crinoids on Vancouver Island. *Uintacrinus socialis*, a stemless crinoid with a large flexible calyx (head), used its long flexible arms to swim through the water in search of food. The spherical calyx is up to 6 cm in diameter and, when complete, the arms could extend to 2 m.

TRACE FOSSILS

Trace fossils indicate different types of behaviour of animals living on the bottom. They can be classified as resting, crawling, or grazing traces, or as feeding or dwelling structures. The three examples illustrated here are grazing traces. In addition to the ethological (behavioural) classification, each trace fossil is given an individual scientific name as if it were a separate organism. Trace fossils are quite common in Upper Cretaceous shales of Vancouver Island, but they have not yet been properly studied.

Taphrhelminthopsis (Figure 117). This trace fossil with the 200-dollar name is a flat, meandering trail with prominent lateral ridges; it appears to be the grazing trail of a large gastropod.

Zoophycus (Figure 117). These organic structures, which look like swirls of a mop in sand, have been variably interpreted as an animal or plant body fossil, or as trace fossils produced by a worm-like animal foraging for organic material in sediment. The name *Zoophycus* reflects this uncertainty—"zoo" meaning animal and "phycus" seaweed. Recent work has proved it to be a trace fossil that is most common in muds deposited in deep water.

Figure 117 The trace fossils *Taphrhelminthopsis* (VIPM 105), Lambert Fm., Collishaw Point, Hornby Island and *Zoophycus* (VIPM 106), Trent River Fm., west side of Denman Island.

Scolicia (Figure 118). The Trent River Formation exposed on the west side of Denman Island contains very few body fossils, but some of the sandstone beds are crowded with startlingly fresh-looking trace fossils. One of these traces consists of meandering trails 3 cm across that comprise a median ribbed axis flanked by striated lateral parts. This is clearly a feeding trail of an organism that fed on the organic matter in the sand. The trail maker cannot be identified with certainty, but it was probably a heart urchin.

Figure 118 The trace fossil *Scolicia* (School of Earth and Ocean Sciences, University of Victoria). Slab is 40 cm across. Trent River Fm., Denman Island.

FISHES

Teeth (Figure 119). A fossil shark tooth with its gleaming brown enamel and its edges still keen after 70 million years is a vivid record of these top-level marine predators. Two types of sharks are represented as fossils in Cretaceous rocks of Vancouver Island—the lamnoids and the hexanchoids. The lamnoids are active sharks with fusiform bodies, such as porbeagle, thresher, mako, and great white sharks. Their teeth, which are shed regularly during life, consist of a broad bony base and a sharp sabre-like central tooth that may be flanked by smaller cusps. It is not clear to which lamnoid sharks these fossil teeth belong. Only a few teeth can be identified to the species level. The hooked triangular tooth with fine "steak knife" serrations on the edge belongs to the lamnoid shark *Squalicorax kaupi*.

The hexanchoids are a group of primitive sharks ("cow sharks") that bear teeth with multiple, sharp, triangular cusps. Living hexanchoids include *Hexanchus griseus*, the six-gill shark, which is a regular visitor to Lambert Channel off Hornby Island. Teeth of its distant relative *Notidanodon lanceolatus* are found as fossils in the Lambert Formation on the same island.

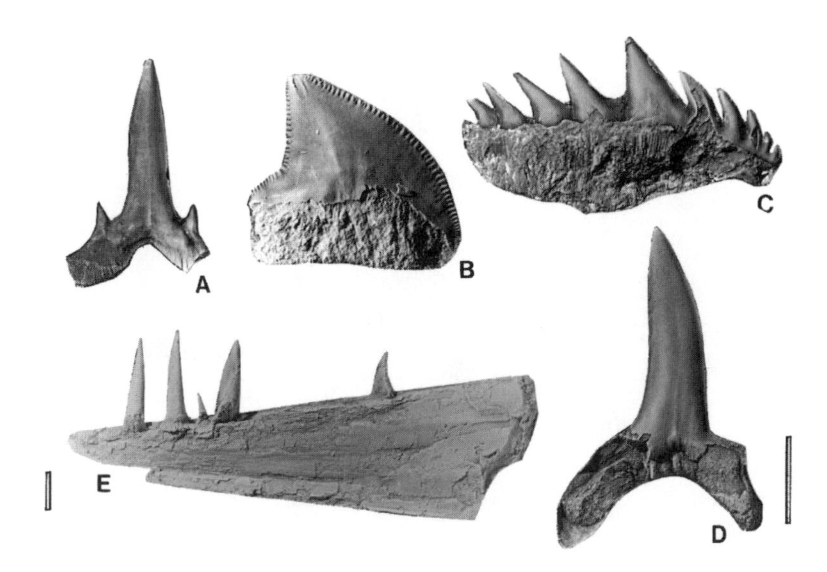

Figure 119 (A-D) Various shark teeth. (A) Lamnoid tooth (CDM 017), Trent River Fm., Oyster River. (B) Lamnoid tooth *Squalicorax kaupi* (VIPM 107), Haslam Fm., Brannan Lake. (C) Hexanchoid tooth *Notidanodon lanceolatus* (VIPM 108), Lambert Fm., Collishaw Point, Hornby Island. (D) Lamnoid tooth (VIPM 109), Oyster Bay Fm., Shelter Point. (E) Lower jaw of an enchodontid fish (VIPM 110), Lambert Fm., Collishaw Point, Hornby Island.

Bones (Figures 119 and 120). Sharks possess cartilaginous skeletons that do not fossilize easily. Only the vertebrae of sharks are well calcified in order to withstand the forces associated with vigorous swimming. These vertebrae (which probably belong to lamnoid sharks) are moderately common in concretions in the lower Trent River Formation exposed on the Puntledge River. They have a curious shape. Each consists of four striated cones arranged with their apices together to produce the form of the letter X.

The successful fishes of the present day have a good fossil record because, as any piscivore can attest, they are bony. Most disarticulated bone cannot be identified, but we have collected a large jaw with long, widely separated teeth from the Lambert Formation on Hornby Island that can be identified as belonging to a member of the family Enchodontidae—large-headed bony fishes with greatly elongated

Figure 120 Shark vertebrae (VIPM 111), Trent River Fm., Browns River and an enchodontid fish gill cover (VIPM 112), Haslam Fm., Brannan Lake.

Figure 121 Scales of the fish *Chicolepis* (VIPM 114), Trent River Fm., Northwest Bay.

teeth (hence the family name, "lance tooth"), which today occupy the deep sea.

Scales (Figures 120 and 121). Modern bony fishes are covered with a flexible armour of overlapping scales. Similar fossil scales are readily preserved in the rock record. In Cretaceous shale units on Vancouver Island, small scales are conspicuous because they commonly have a bright blue or purple sheen. The concentric pattern of growth lines, like a fingerprint, can be seen with the aid of a hand lens. Large square scales with finely etched microsculpture from the Haslam Formation at Brannan Lake belong to the genus *Chicolepis*, which seems to be related to the tarpon, a modern salmon-shaped fish that has not changed much since the Cretaceous. An oblong scale-like element with fine ridges radiating from the edge is the gill cover of an enchodontid fish.

Protosphyraena (Figure 122). This peculiar bony fossil consists of a pointed shaft 20 cm long with rounded and sharp-edged knobs on one side. Although it consists of solid bone, it is actually a fused pectoral fin of the predaceous barracuda-like fish *Protosphyraena*. These paired fins, located behind the head, would have been able to slam sideways and forward to kill or stun other fish in a school. The entire fin would have measured half a metre in length, and the fish itself might have been 3 m long. *Protosphyraena* is moderately common in the Upper Cretaceous chalk of Kansas, where its fossil record consists mainly of its solid pectoral fin. It is rare in Canada and it has not previously been documented from the west coast Cretaceous.

Figure 122 Pectoral fin of the fish *Protosphyraena* (CDM), Trent River Fm., Trent River.

Coprolites (Figure 123). Peculiar, corrugated, torpedo-shaped fossils from the English Upper Cretaceous Chalk sparked scientific curiosity as early as the 1700s. First described as petrified larch cones, their true nature as fossil feces was established in 1829 by the Rev. William Buckland, Professor of Geology at Oxford University. Buckland coined the name coprolite (Greek for "dung rock") for these fossils. Fish coprolites can provide considerable information—not only about the dietary habit of the fish, but also about the shape and nature of the lower intestine. Moreover, some small Cretaceous fossils from Vancouver Island are known only because these animals had the misfortune to be swallowed by fishes and were preserved in coprolites. The cockroach wing (see Figure 104) is a good example of this type of preservation.

Figure 123 Coprolites of a bony fish (VIPM 115), Lambert Fm., Collishaw Point, Hornby Island and a shark (VIPM 116), Haslam Fm., Brannan Lake.

The striated lenticular coprolite from the Haslam Formation shown in Figure 123 consists of overlapping spirals. It was formed within the lower intestine of a Cretaceous shark. Feces of living sharks, such as dogfish, consist of a long ribbon that is spirally coiled in the lower colon, where it hardens before it is voided. Other kinds of coprolites are fairly common in concretions in the Haslam Formation. These consist of diffuse smears of fish scales and fish spines—the undigestible remains of a meal of fish-eating fishes.

Hornby Island Shark Teeth (Figure 124). Shark teeth are eagerly sought-after fossils and, although they occur widely, are never common in the rocks of Vancouver Island. A diligent collector, however, will generally be rewarded by one or two gleaming specimens after a couple of hours search at most Cretaceous fossil localities. Kurt Morrison, amateur paleontologist and goldsmith from Hornby Island, is not satisfied with a few specimens. He has documented an astonishingly rich assemblage of shark teeth in a 10-cm-thick grey-green shale unit of the Lambert Formation at Collishaw Point. Each tooth has been freed from its shale matrix—necessary because the inner and outer sides differ and because teeth vary according to their position in the mouth. More than 600 isolated teeth can be assigned to 18 species belonging to eight types of sharks (frill, cow, bramble, dogfish, saw, angel, horn,

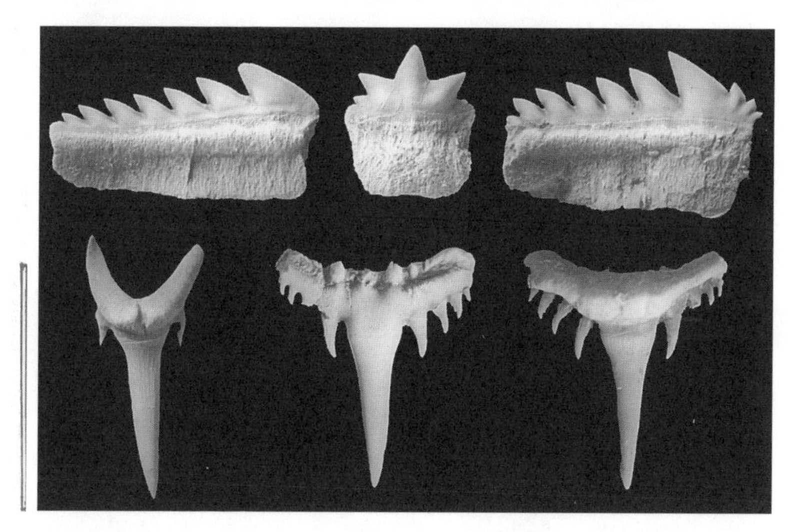

Figure 124 Shark teeth from Kurt Morrison Collection. (Top) the cow sharks *Hexanchus microdon* (two teeth) and *Heptranchias* sp., (Bottom) the sand tiger shark *Carcharias* sp., and the horn shark *Synechodus* sp. (both sides of one tooth). All from Lambert Fm., Collishaw Point, Hornby Island (photos by K. Morrison and B. Hessin).

and sand tiger sharks). This is the most diverse deep-water shark assemblage yet discovered in Upper Cretaceous rocks anywhere in the world. Here we show four species from Kurt's collection—cow sharks (*Hexanchus microdon* and *Heptranchias* sp.), sand tiger shark (*Carcharias* sp.) and a horn shark (*Synechodus* sp.).

MARINE REPTILES

Elasmosaurs (Figures 125 to 129). The famly Elasmosauridae was a group of Jurassic and Cretaceous marine reptiles having a streamlined body, a short tail, two pairs of paddle-like limbs, a long flexible neck, and a small head with long interlocking teeth used to capture and trap fish. These animals, appropriately known as swan lizards, reached impressive dimensions—up to 12 m in length, half of which was neck. Unlike most other reptiles, living or extinct, elasmosaurs must have been viviparous (that is, giving birth to live young) because their bulk precluded their laying eggs on land.

A reconstruction of a living elasmosaur is familiar to most people because these fossil animals have served as unwitting models for

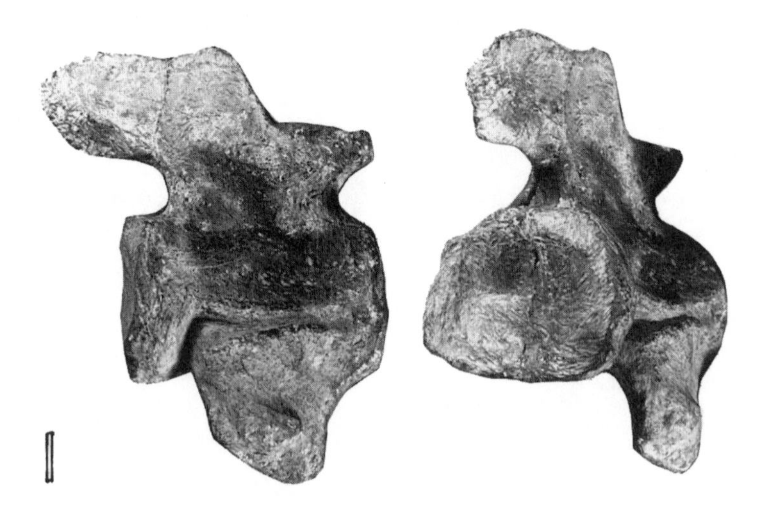

Figure 125 Neck vertebra of the Puntledge elasmosaur in side and oblique view (CDM). Note high backswept neural spine, lateral processes, and the neural canal. Trent River Fm., Puntledge River.

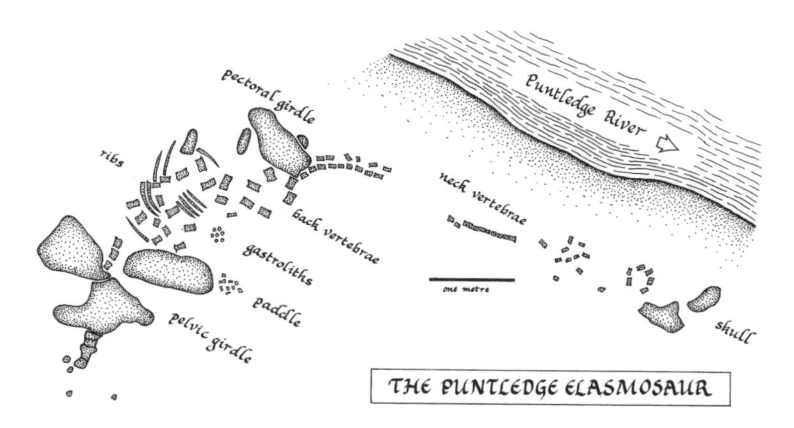

Figure 126 Plan of the bones of the elasmosaur excavated by volunteers for the Courtenay Museum in the spring of 1991 from hard siltstones of the Trent River Formation exposed along the Puntledge River.

drawings of the imaginary monsters supposedly inhabiting lakes such as Loch Ness and Lake Okanagan, and marine waters such as Cadboro Bay in Victoria. Notwithstanding fanciful headlines in the tabloids (often furtively read at supermarket checkout counters!), elasmosaurs became extinct at the end of the Cretaceous.

The first west coast elasmosaur was discovered by Mike Trask of Courtenay in shales exposed on the Puntledge River. An excavation by volunteers supervised by a paleontologist in spring 1991, resulted in the recovery of eighty-two bones—including sixty-six vertebrae, nine ribs, large concretions containing pelvic (hip) and pectoral (chest) girdles, and at least one paddle-like limb, as well as two concretions with the skull and jaw (Figure 126). All the bones of the Puntledge elasmosaur have now been prepared and cleaned. Some of the vertebrae proved to be in excellent condition with complete neural spines and canals, and lateral processes (Figure 125). Unfortunately, the skull was crushed beyond repair, but the jaw is well preserved and, with a length of 50 cm, unusually large for an elasmosaur. The teeth are circular in cross-section, long, slightly curved, and etched with very fine lines (Figure 129). A cast of the reconstructed skeleton of the Puntledge elasmosaur now hovers over the displays at the Courtenay Museum (Figure 127).

Figure 127 Reconstructed cast of the Puntledge elasmosaur suspended from the ceiling of the Courtenay Museum.

Partial skeletons of elasmosaurs have also been collected from shales of the Trent River formation exposed on Trent River and on Englishman River. Neither is as complete as the Puntledge elasmosaur, but the Englishman River specimen is of considerable interest because within its ribcage is preserved a mass of highly polished stones of volcanic rock (Figure 128). These are gastroliths, which traditionally have been interpreted to be gizzard stones that aided in the maceration of the fish that were the main prey. Some elasmosaurs contain hundreds of gastroliths with a total weight of 10 kg. Because it is unlikely that elasmosaurs had gizzards, the gastroliths were probably stomach stones that served as ballast to improve stability of the elasmosaur in water and to counteract the buoyancy of the large air-filled lungs during dives. The stomach stones of living crocodiles serve these functions.

Mosasaurs (Figures 129 and 130). Mosasaurs were large, slender, crocodile-like marine reptiles distantly related to the living monitor lizard. These sea lizards used their long tails to scull through the water and their peg-like teeth to kill ammonites, nautiloids, and fish. The illustrated specimens of the ammonite *Pachydiscus* (Figure 130) and the nautiloid *Eutrephoceras* (see Figure 75) from the Lambert Formation bear the unmistakable teeth marks of a sea lizard. Mosasaurs are distributed world-wide in Upper Cretaceous rocks and, like almost all other large animals, they became extinct at the mass

Figure 128 Gastroliths of highly polished volcanic pebbles from the ribcage of an elasmosaur (VIPM). Trent River Fm., Englishman River.

Figure 129 A Cretaceous dentary. Two mosasaur teeth 2.5 cm long (K. Morrison Collection), Lambert Fm., Collishaw Point, Hornby Island. A theropod dinosaur tooth 1.0 cm long in lateral and front view (CDM), Trent River Fm., Trent River. A 10 cm long tooth from the Puntledge elasmosaur (CDM), Trent River Fm., Puntledge River.

Figure 130 Skull of the mosasaur *Tylosaurus* sp. (VIPM 117) and its bite marks on the ammonite *Pachydiscus suciaensis* (CDM). Both Lambert Fm., Collishaw Point, Hornby Island.

extinction at the end of the Cretaceous. A partial skull of the mosasaur *Tylosaurus* has been recovered from the Lambert Formation on Hornby Island, and eight vertebrae of another mosasaur came from the lower Trent River Formation on the Puntledge River.

FLYING REPTILES

Pterosaur (Figure 131). A thin-walled and hollow bone, 25 mm long, is the sole fossil evidence that pterosaurs—flying reptiles—were living on the west coast during the Cretaceous. It was collected and prepared by Kurt Morrison from the productive "shark teeth" shale at Collishaw Point on Hornby Island. The bone appears to be a wing metacarpal, the stoutest of four metacarpals that form the bones of the palm of the hand. The rounded processes at the distal end (to the left) would articulate with the greatly lengthened wing finger that supports the wing membrane. The proximal end (to the right) would articulate with the wrist bones. The deep longitudinal groove suggests the presence of strong ligaments that would have controlled movement of the wing.

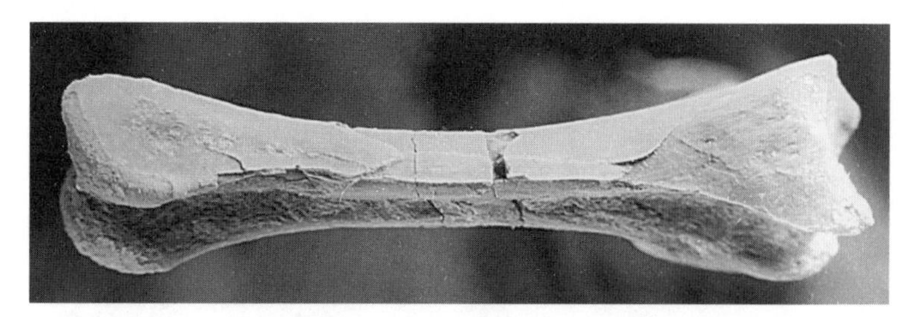

Figure 131 A pterosaur hand bone 2.5 cm long (K. Morrison Collection). Lambert Fm., Collishaw Point, Hornby Island (photo by K. Morrison and B. Hessin).

However, little is definitive when it comes to identification of small isolated fossil bones. When Dr. Peter Wellnhofer from the Bavarian Collections of Paleontology in Munich (and a pterosaur specialist) examined the photograph above, he suggested that it was "probably a finger or toe bone; seemingly not from a pterosaur, but possibly from a small dinosaur." As usual, more material is needed.

DINOSAURS

Theropod (Figure 129). Although bones of dinosaurs have not been located on Vancouver Island, dinosaurs undoubtedly lived in the forested lands that surrounded the Georgia Basin during the Cretaceous. A centimetre-long triangular tooth found in marine shales of the Trent River Formation on Trent River is a theropod tooth—the first evidence that dinosaurs lived west of the Rocky Mountains. This tooth has a sharp front edge and its sides are scored by irregular grooves. Theropods, like all dinosaurs, shed their teeth throughout life, but this tooth was probably not shed because it appears to include bony material at its base. How did this tooth end up at the bottom of the deep sea? One could speculate about a small theropod carcass floating down to the sea, where it was ingested by marine reptiles. The teeth were excreted and fell to the deep mud bottom.

The Theropoda is the group of bipedal carnivorous dinosaurs that includes the ceratosaurs, carnosaurs, tyrannosaurs, and coelurosaurs. Some paleontologists maintain that theropods are still with us—as birds, which unquestionably were derived from the coelurosaurs in the Jurassic.

Figure 132 The cycad *Pseudoctenis* sp. (RBCM 91810), Comox Fm., Cumberland and a reproductive organ of the cycadeoid *Cycadeoidea* (VIPM 118), Lambert Fm., Collishaw Point, Hornby Island.

CYCADS AND CYCADEOIDS

Cycads, also called sago palms, and their extinct relatives, the cycadeoids, have short, thick, barrel-shaped trunks surmounted with long, stiff, fern-like leaves.

Pseudoctenis (Figure 132). The name *Pseudoctenis* is used for cycad leaves that have long straight leaflets arising from the axis at a high angle. The veins extend unbranched to the tip of the leaflet. *Pseudoctenis* sp. occurs widely in the Comox Formation in the Cumberland area, in the Protection Formation near Nanaimo, and in the Suquash Formation near Port Hardy.

Cycadeoidea (Figure 132). The cycadeoids, which, unlike their cousins the cycads, did not survive into the Cenozoic, are represented in the Lambert Formation by a few remarkable cones containing the reproductive organs of the genus *Cycadeoidea*. This plant consisted of cycad-type leaves emanating from a squat stem with numerous cones on its sides. For many years, these cones were thought to be flowers and, therefore, the cycadeoids were considered likely ancestors to the angiosperms. The origin of the flowering plants is still problematic, but it is clear that they did not arise from the cycadeoids. The cycadeoid cone, about the size and shape of a light bulb, consists of long, thin, densely packed cylindrical scales that protected and enclosed the seeds.

MAIDENHAIR TREE

Ginkgo (Figure 133). The term "living fossil" was first coined by Charles Darwin for the maidenhair tree *Ginkgo biloba*, the sole living species of an ancient group of gymnosperms that dates back to the Triassic. This term, almost an oxymoron, refers to living species that have persisted with little change for long spans of geologic time. G. *biloba* grows wild only in a small area in western China, but because of its hardiness and

Figure 133 The maidenhair tree *Ginkgo dawsoni* (GSC 5683), Suquash Fm., Port McNeill and the conifer *Protophyllocladus* sp. (VIPM 119), Protection Fm., Nanaimo.

its attractive pale green foliage borne on graceful limbs, it has become a common ornamental tree that grows well in temperate regions, even in the polluted centres of European and North American cities. The characteristic fan-shaped leathery leaf of *Ginkgo dawsoni* has been collected from the Suquash Formation at Port McNeill. This species celebrates the most eminent scientist of late Victorian Canada—Rev. J. William Dawson, geologist, paleontologist, principal of McGill University in Montreal, and arch-foe of Charles Darwin's theory of evolution by natural selection. Rev. Dawson was also the father of George Mercer Dawson, who made important contributions to the geology, paleontology, botany, and ethnology of British Columbia in the late 1800s.

CONIFERS

Protophyllocladus (Figure 133). Flattened, densely striate, leaf-like branches with lobed margins occur in virtually all of the plant-bearing Cretaceous formations on Vancouver Island. Identified as *Protophyllocladus* sp., these plants are classified with *Phyllocladus* in the family Podocarpaceae—a group of conifers that is largely restricted to areas south of the equator, but also grown as house plants (for

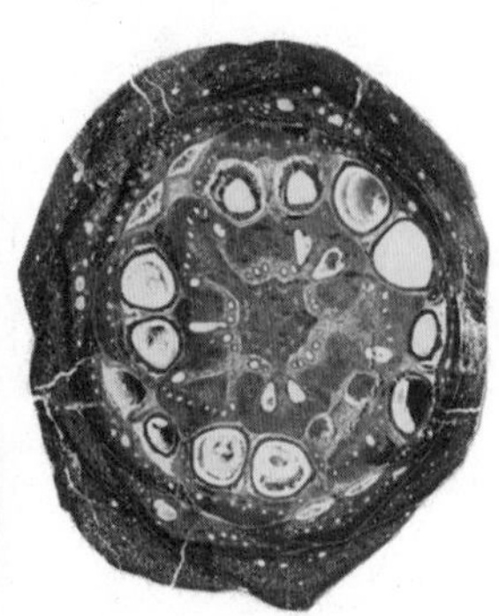

Figure 134 The conifers *Elatocladus* sp. (VIPM 120, 121), Protection Fm., Nanaimo and Lambert Fm., Collishaw Point, Hornby and *Pinus* sp., a cross-section of a permineralized pine cone (CDM), Haslam Fm., Brannan Lake.

example, the Buddhist pine). The presence of this latter group in the Cretaceous of North America has not yet been demonstrated. We consider *Protophyllocladus* to be a group of extinct conifers.

Elatocladus (Figure 134). The redwood family Taxodiaceae bears flattened foliage with needles arranged in two rows and ovoid cones borne on long stalks. Only a few genera survive to the present time, mostly represented by a single species each. These include the redwood (*Sequoia*) of California, the dawn redwood (*Metasequoia*) of northeastern Asia, and the bald cypress (*Taxodium*) of southeastern United States. Unless well-preserved cones are found, fossil foliage, such as that from the Vancouver Island Cretaceous, cannot be assigned confidently to any one of these genera. Instead, we use the name *Elatocladus* for the redwood foliage that occurs widely in the Comox, Protection, and Suquash formations.

Pinus (Figure 134). A well-preserved permineralized conifer cone from the Haslam Formation at Brannan Lake probably belongs to the pine genus *Pinus*. A polished cross-section shows that it bears long, thick, woody, overlapping cone scales, each covering a pair of seeds. This unique fossil is important because it constitutes one of the earliest records of *Pinus*.

Pinaceae (Figure 141). Another extraordinarily well-preserved conifer cone comes from a sandy concretion in the Oyster Bay Formation at Shelter Point. Oblong and cylindrical in shape, this is an ovulate (female) cone displaying two rows of seeds. It is similar to both living *Abies* (true fir) and *Picea* (spruce), but because neither of these conifers has been firmly documented in the Late Cretaceous, we identify this cone only to the family level.

Conifer wood (Figure 135). Well-preserved permineralized conifer driftwood is common in concretions from the Lambert Formation on Hornby Island. Most pieces have been bored by Cretaceous teredo bivalves. The sharply expressed annular rings, clearly seen in the polished cross-section, indicate a pronounced seasonality in this area during the Late Cretaceous.

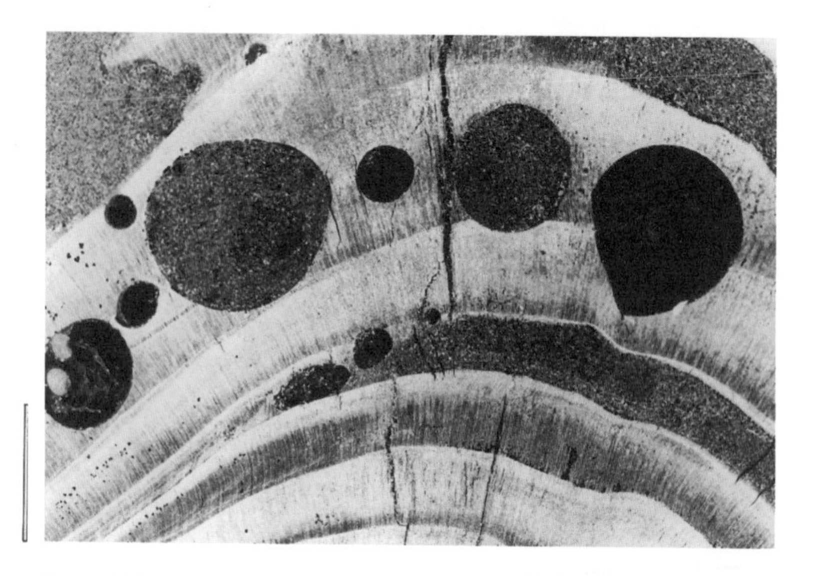

Figure 135 Conifer wood bored by teredo bivalves (VIPM 122), Lambert Fm., Collishaw Point, Hornby Island.

MONOCOT ANGIOSPERMS

Sabalites (Figure 136). Fragments of sabaloid palm leaf-stalks are rare in concretions in the Lambert Formation on Hornby Island, but it is remarkable that these fossils are there at all. These long, flat leaf-stalks, about 5 cm across, are armed with backwardly hooked spines. The leaf-stalks branch into numerous rays arranged like a fan. *Sabalites* is well known at a number of Late Cretaceous tropical and subtropical sites. The palmetto palm (*Sabal*) of southeastern United States and the sentinel palm (*Washingtonia*) of California are living members of this group of sabaloid palms.

Palm wood (Figure 137). A couple of fossil palm logs, possibly *Sabalites*, have been collected from the Lambert Formation on Hornby Island. A polished cross-section shows densely packed, triangular leaf bases. The vascular bundles of monocots are scattered throughout the wood, not in discrete cylinders as in dicot angiosperm wood.

Figure 136 The leaf stalk and base of frond of the sabaloid palm *Sabalites* sp. (VIPM 123, 124), Lambert Fm., Collishaw Point, Hornby Island.

Figure 137 Cross-section of palm wood (VIPM 125) and angiosperm hardwood (VIPM 126), Lambert Fm., Collishaw Point, Hornby Island.

DICOT ANGIOSPERMS

Dicot angiosperm wood (Figure 137). The fossil record is an informative mosaic of the familiar and the strange. A 70-million-year-old permineralized log from the Lambert Formation on Hornby Island is essentially identical to recent hardwoods. A polished cross-section of this angiosperm log shows rays emanating from the central pith. There is a pronounced colour difference between an inner, darker xylem (wood proper—the water-conducting tissue) and an outer, lighter phloem (the food-conductive tissue) underneath the bark, which is not preserved.

Zizyphus? (Figure 138). This distinctive ovate leaf with three strong veins that extend from the base to the tip is quite common in all Upper Cretaceous plant-bearing formations on Vancouver Island. The name is inappropriate, however, as these fossil leaves appear to be entirely unrelated to living *Zizyphus*, the jujube tree. Therefore we have queried the name.

Dryophyllum (Figures 138 and 139). The abundance of this type of leaf in most Cretaceous plant-bearing formations on Vancouver Island, as well as in rocks of the same age in Alberta, Wyoming, and Utah, indicates that it must have belonged to one of the most important trees

Figure 138 The angiosperm leaves *Zizyphus?* sp. (VIPM 127) and *Dryophyllum* sp. (VIPM 128, 129). All from Trent River Fm., confluence of Browns and Puntledge rivers.

Figure 139 The angiosperm leaves *Dryophyllum* sp. (VIPM 130), *Laurophyllum* sp. (VIPM 131), and *Platanus* sp. (VIPM 132). All from Trent River Fm., confluence of Browns and Puntledge rivers.

of the forest canopy. Leaves of *Dryophyllum* have a strong, straight midrib and offset secondaries that weaken toward the margin, which is irregularly toothed. The shape is highly variable from wide elliptical to narrow lanceolate. *Dryophyllum* is similar to leaves of oak (*Quercus*), ash (*Fraxinus*), and willow (*Salix*), but an assignment to any living genus would probably be wrong. The broad application of this name is appropriate because *Dryophyllum* is Greek for "tree leaf."

Laurophyllum (Figure 139). Leaves of this type are elliptical, with a straight primary vein and a few strong secondary veins all arising from the base. Other secondaries come off the primary at high angles. *Laurophyllum* may well be the leaf of a member of the Lauraceae, the laurel family, an ancient angiosperm group dating back to the Late Cretaceous.

Platanus (Figures 139 and 140). Leaves of the sycamore or plane tree, the family Platanaceae, are moderately common in Cretaceous rocks of Vancouver Island. These leaves have coarse marginal teeth and may be trilobed. The two secondaries come off the base of the primary rib at slightly different points. These leaves are confidently assigned to *Platanus*, probably the oldest genus of flowering plant still living. At the present time, sycamores are common trees in moist lowland forests in warm temperate regions.

Ternstroemites (Figure 140). These leaves are elliptical in outline and have a stout midrib. The secondaries come off the midrib at a right angle, but they fade rapidly and fail to reach the margin. The name *Ternstroemites* is used for this leaf type, which is similar to leaves of the living walnut (*Jugulans*).

Figure 140 The angiosperm leaves *Platanus* sp. (VIPM 133) and *Ternstroemites* sp. (VIPM 134). Both from Trent River Fm., confluence of Browns and Puntledge rivers.

Fruit and nuts (Figure 141). Collectors rarely find permineralized angiosperm fruit and nuts in Cretaceous rocks of Vancouver Island (or anywhere else, for that matter), but they are potentially so significant that considerable effort should be expended to find additional specimens. To study these fossils properly, they must be cut and polished. An unusual fruit from the Haslam Formation bears multiple double

Figure 141 Angiosperm fruit, possibly Magnoliaceae (VIPM 135), Haslam Fm., Brannan Lake; a large seed (VIPM 136), Oyster Bay Fm., Shelter Point; and a cone of the family Pinaceae (VIPM 154), Oyster Bay Fm., Shelter Point.

rows of seeds borne within the flesh, now turned to stone. The surface of this fruit is eroded, thus exposing the interior. Similar many-seeded fruits are produced by various members of the Magnoliaceae—a family that dates back to the mid-Cretaceous.

THE CRANBERRY ARMS SITE

On a hot day in early August 1996, bulldozer operator John Bell was busy moving large chunks of blasted sandstone along the approach to the new ferry terminal at Duke Point, south of Nanaimo. He spotted something peculiar on the underside of a massive piece of grey rock. It was not a fracture—the closely spaced lines extending from a central axis were more reminiscent of a gigantic feather. It was definitely

Figure 142 The palm *Phoenicites imperialis* photographed in front of the Cranberry Arms Inn. Rock face measures 1.5 m across. This specimen is now at Malaspina College, Nanaimo (photo by Maggie McColl).

unusual. So, instead of adding this rock to those destined for crushing, he lifted the basket of the bulldozer up 3 metres and set the rock at the edge of the roadcut, right in front of the Cranberry Arms Inn. A stunning plant fossil could then be seen—a palm frond nearly 2 metres long, possibly the largest fossil leaf ever found in Canada.

Other plant fossils were also present: angiosperm leaves with the veins in relief, conifer foliage, delicate fern fronds, and, most surprising of all, exquisitely preserved flowers. The press was alerted and a spate of articles in Nanaimo and Victoria newspapers followed (one predictably referred to these plant fossils as "dinosaur salad"). Soon, amateur paleontologists descended on the site. And, in what can only be described as a rescue mission, fossiliferous blocks were snatched, literally in front of working bulldozers, to prevent their being crushed.

Now, little remains to be collected at the site. Only a vertical rock wall; fortunately, substantial collections of these important plant fossils are held by various amateur paleontologists on Vancouver Island and by a few museums. The original palm frond discovery at the Cranberry Arms site is now at Malaspina College in Nanaimo. We show a few palms, ferns, flowers, and leaves of this important Late Cretaceous fossil flora.

Palm (Figure 142). Fronds of the palm *Phoenicites imperialis* at the Cranberry Arms site are extremely long. Each frond consists of a stout, gradually tapering rachis (a main axis of a compound leaf) up to 2 m long that supports the rays of the leaf. The rays form a continuous sheet of 1-cm-wide plications that have the appearance of a fine venetian blind. In life, these leaves must have been leathery and tough. The generic name alludes to a similarity to the living date palm, *Phoenix*. It is unlikely that they are closely related.

Ferns (Figure 143). Ferns are common and well preserved at the Cranberry Arms site. *Asplenium* carries small pointed and serrated leaflets that are slightly offset on the rachis. Each leaflet has a central vein. The name alludes to the plant being a supposed curative for diseases of the spleen. *Coniopteris* has lobate to pointed leaflets closely adhering to the rachis. It is very widespread in Jurassic and Cretaceous rocks of western North America.

Flowers (Figure 144). Flowers do not have much potential for fossilization, but the Cranberry Arms site includes unequivocal fossils of two floral types.

With 1800 living species of perennial herbs, the buttercup family Ranunculaceae is large. It includes columbine, larkspur, clematis,

Figure 143 A Cretaceous fern bouquet of *Asplenium* on the left and a frond of *Coniopteris* on the right (J. Whittles Collection). Protection Fm., Cranberry Arms site.

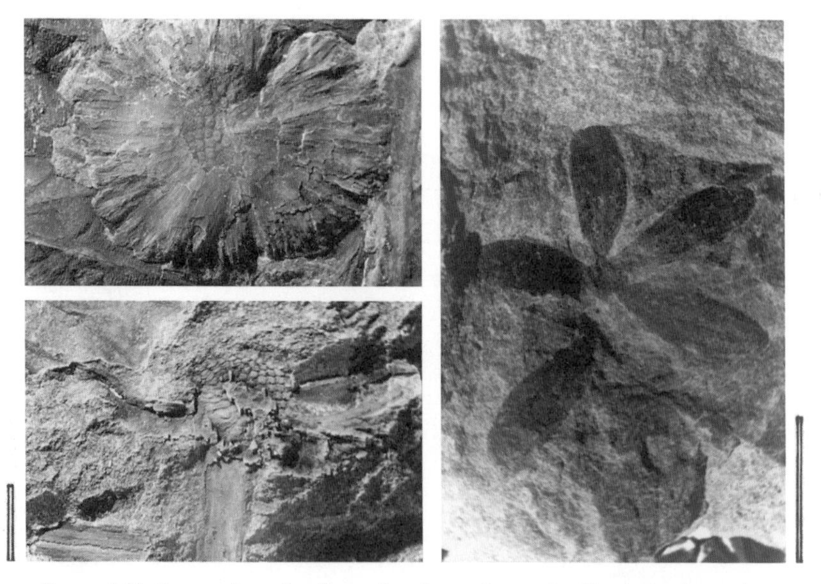

Figure 144 Flowers from the Protection Formation at the Cranberry Arms site: Buttercup family Ranunculaceae, top and side view and *Astronium?*, top view (J. Whittles Collection).

crowfoot, as well as numerous species of the ubiquitous buttercup. This ancient angiosperm family, dating from the Late Cretaceous, is represented by well-preserved flowers at the Cranberry Arms site. In these unlikely fossils, the calyx is held aloft by a stout woody stalk. The petals appear to be numerous and radially arranged. The central part of the flower shows that many stamens and pistils were present. This Cretaceous "buttercup" is preserved both compressed and with the petals splayed out.

The other fossil flower at this locality bears a calyx of five oblong sepals, each with three parallel veins. It is virtually identical to the flowers of *Astronium*, a tropical hardwood of the sumac family—Anacardiaceae, a group that also contains poison ivy and that has a fossil record extending back to the Paleogene. This family has not previously been reported from the Cretaceous. We identify this small flower as *Astronium*?.

Leaves (Figure 145). We can't show all of the different kinds of fossil angiosperm leaves that were found at the Cranberry Arms site. One leaf will have to represent the lot. This attractive leaf shows considerable relief, with the primary and secondary veins prominent. Fossil leaves of this kind have been called *Viburnum*, even though few paleobotanists believe they are the same as the popular garden shrub.

Figure 145 Leaf of the angiosperm *Viburnum* (J. Whittles Collection), Protection Fm., Cranberry Arms site.

CENOZOIC

Cenozoic sandstone and shale units clinging to the Pacific rim between Sooke and Nootka Sound preserve the youngest fossils found on Vancouver Island (with the exception of those occurring in Pleistocene sands and muds). This narrow outcrop belt is rarely more than a few kilometres wide (Figure 146). The older unit, called the Hesquiat Formation, consists of a thousand metres of silty shales interspersed with a few sandstone and conglomerate beds. These sediments accumulated in deep marine waters. The younger unit, the Sooke Formation, is a few hundred metres thick and consists of extremely fossiliferous pebbly sandstones that were deposited in shallow water.

The fossil tusk shells, clams, snails, crabs, and barnacles preserved in these rocks are very similar to those now living around Vancouver Island. Teeth of both lamnoid and hexanchoid sharks occur in the Hesquiat Formation, and teeth and bone of three different kinds of extinct marine mammals are found in the Sooke Formation. Abundant but poorly preserved and carbonized foliage of beech, laurel, magnolia, hickory, spruce, oak, and willow also occur in the Sooke Formation, and a well-preserved pine cone has been described from the Hesquiat Formation. The Hesquiat is late Paleogene in age and the Sooke early Neogene.

Along the beach some 12 km south of Campbell River and within sight of the well-known Upper Cretaceous fossiliferous locality of the Oyster Bay Formation at Shelter Point, a succession of concretionary shales and sandstones is exposed. These rocks seem to be run-of-the-mill marine Cretaceous strata, as seen elsewhere in the Comox Basin. Indeed, the Geological Survey of Canada mapped these rocks as Upper Cretaceous. However, the fossils demonstrate an age younger than Cretaceous. These beds are Paleogene; the only marine rocks of that age exposed anywhere on the east coast of Vancouver Island. These rocks have not been formally named. We refer to them as the Via Appia Beds (after the Appian Way, the name of the road access off the Island Highway). Although these rocks are marine, a concretionary unit includes extraordinarily well-preserved seeds and fruits of flowering plants.

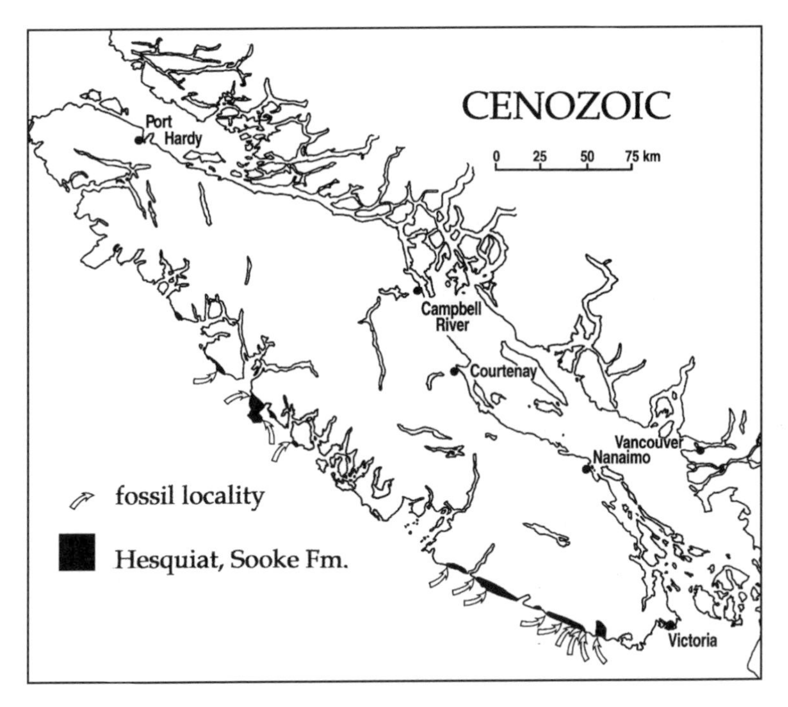

Figure 146 Distribution of Cenozoic rocks on Vancouver Island and fossil localities. The Via Appia Beds south of Campbell River and the Chuckanut Formation on Tumbo Island off Saturna Island are not shown on this map.

Cross-bedded sandstones of the Chuckanut Formation form the bedrock of Tumbo Island off the north coast of Saturna Island. They are about the same age as the Hesquiat Formation and Via Appia Beds, but they are non-marine and deposited by laterally shifting rivers. These sandstones contain fossil pollen, but they lack larger fossils.

PUBLISHED REFERENCES—Cenozoic

ROCKS—Cameron (1980); Jeletzky (1975); Muller, Cameron, and Northcote (1981); Mustard and Rouse (1994).

FOSSILS—Clark and Arnold (1923), Merriam (1899), McKenzie McAnally (1996), Rathburn (1926), Russell (1968), Waldman (1971).

SCAPHOPODS

Dentalium (Figure 147). Scaphopods are represented in the Hesquiat Formation by large, almost cylindrical tubes covered by net-like sculpture of faint longitudinal ribs and fine growth lines. Like living members of the genus, *Dentalium* sp. lived in deep waters.

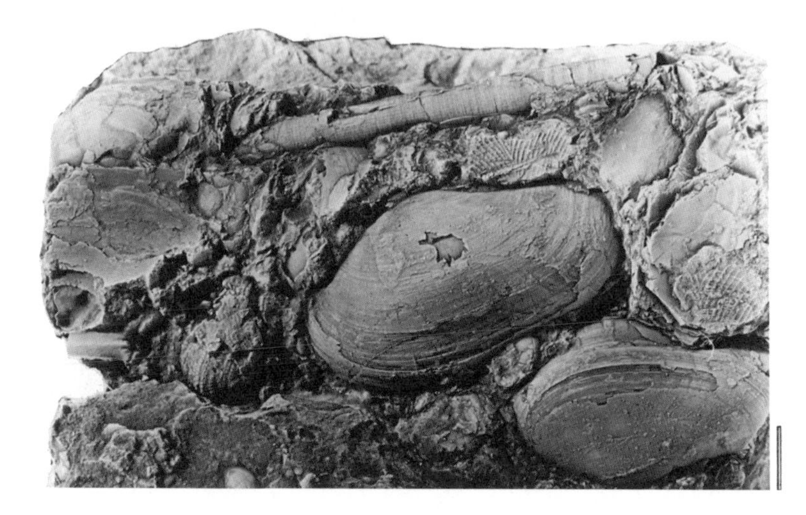

Figure 147 Shell bed with the bivalve *Yoldia blakeleyensis* and the scaphopod *Dentalium* sp. (GSC 43217), Hesquiat Fm., Hesquiat Peninsula.

BIVALVES

Yoldia (Figure 147). Most living infaunal bivalves are filter-feeders that strain minute food particles from the water. *Yoldia*, however, which lives partially buried in sand with its posterior siphons exposed, is a surface-deposit-feeder that gropes for food with its modified gills. This bivalve has a cosmopolitan distribution and a fossil record that extends back to the Cretaceous. *Yoldia blakeleyensis* is a large, smooth, rectangular, and thin-shelled bivalve in the Hesquiat Formation.

Acila (Figure 148). The striking divaricate sculpture of this oval bivalve cuts across the concentric growth lines. The sculpture consists of sets of ribs that form nested chevrons pointing to the umbo. *Acila* is a shallow

Figure 148 The bivalve *Acila gettysburgensis* (GSC 43083) and the crab *Ranidina naselensis* (GSC 43124). Both from Hesquiat Fm., Hesquiat Peninsula.

mud-burrower with a morphology that has remained essentially the same since the Cretaceous. A. *gettysburgensis* from the Paleogene Hesquiat Formation is merely larger than A. *shumardi* from the Cretaceous Haslam and Trent River formations (see Figure 65).

Chlamys (Figure 153). Being the ubiquitous logo of a well-known oil company, the scallop is probably the most familiar of all bivalves. The features that differentiate *Chlamys* from other scallops are its hinge ears of different sizes—the larger anterior ear being undercut by a byssal notch—and its having strong radial ribbing. *Chlamys* was a suspension-feeder that lived byssally attached to floating seaweed.

Spisula (Figure 149). Surf clams are among the largest of the shallow marine infaunal bivalves, reaching lengths of 20 cm or more. *Spisula sookensis* is broadly triangular and essentially featureless on the exterior, even lacking conspicuous growth lines. On the interior, an elevated spoon-shaped structure beneath the umbo serves as a receptacle for the ligament. *Spisula* is probably the most common bivalve in the Sooke Formation.

Mytilus (Figure 149). Not much needs to be said about these familiar mussels which, at the present time, cover almost every hard surface

Figure 149 (Top): the bivalves *Spisula sookensis* (VIPM 137) and *Mytilus mathewsonii* (VIPM 138). (Bottom): the bivalve *Ensis* sp. (VIPM 139) and the hexacoral *Siderastrea vancouverensis* (RBCM). All from Sooke Fm., near mouth of Muir Creek.

exposed in the intertidal zone. They are attached by tough byssal threads. M. *mathewsonii*, from the Sooke Formation, is not quite as slender as the living blue mussel M. *edulis*, but it is identical otherwise.

Ensis (Figure 149). Razor clams live in deep vertical burrows in shallow sands, where they are active filter-feeders. In outline, E*nsis* sp. is rectangular, four times as wide as high, with valves that gape at both ends. Razor clams are uncommon in the Sooke Formation.

HEXACORALS

Siderastrea (Figure 149). Because the Sooke Formation sandstones were deposited in cool temperate waters, corals are rare fossils in this unit. A few fragile, wafer-thin colonial hexacorals assigned to *Siderastrea vancouverensis* encrust some clam shells. Each individual corallite is six-sided with irregular crinkled septae.

GASTROPODS

Polinices (Figure 150). A fossil moon snail gives little indication of the most characteristic feature of a living moon snail—the enormous fleshy foot that almost envelops the rotund shell. But there is little doubt that *Polinices victoriana* from the Sooke Formation belongs to this group of carnivorous snails. Moon snails burrow within the sand and use their extendable radula to drill holes in clams, oysters, and other snails. The shell of *Polinices* consists of a few small whorls followed by a single enlarged whorl.

Figure 150 The gastropods *Polinices victoriana* (VIPM 140, 141), *Acmaea* sp. (VIPM 142), and *Megathura vancouverensis* (RBCM). All from Sooke Fm., near mouth of Muir Creek.

Acmaea (Figure 150). This particular limpet has a slightly asymmetric, oval, cap-shaped shell with vague radial ribs and sharp, closely spaced growth lines. Living limpets of this genus feed on coralline algae. *Acmaea* sp. from the Sooke Formation probably had the same food source.

Megathura (Figure 150). This keyhole limpet is marked by a characteristic star-burst pattern of fine ridges around an apical hole. In life, the exhalent siphon would have emerged through this perforation. This snail probably grazed on algae in the subtidal zone. *Megathura vancouverensis* is a rare fossil in the Sooke Formation, where most of the snails are carnivores, not herbivores.

Levifusus (Figure 151). The intertidal zone around Vancouver Island is now patrolled by a diverse group of predatory snails, the whelks. They perform their grim work by drilling neat circular holes in the shells of barnacles and mussels and then extracting the flesh of their prey with their extendible radula. Closely similar snails are very common in the Sooke Formation. *Levifusus acuminatum*, the dominant whelk in the Sooke Formation, is spindle-shaped, with up to five whorls and two rows of elongate nodes connected by low ridges.

Figure 151 The gastropod *Levifusus acuminatum* (VIPM 143, 144). Both from Sooke Fm., near mouth of Muir Creek.

CRABS

Ranidina (Figure 148). This raninid crab bears an oval carapace with two pairs of short spines on the front margin and a pair of widely divergent spike-like spines on the lateral margin. The surface is coarsely pitted. Like other raninid crabs, its front claws are weak. R. *naselensis* occurs in the Hesquiat Formation at Hesquiat Point and in rocks of the same age in Washington State.

Zanthopsis (Figures 152 and 153). Well-preserved fossil crabs are common in concretions from the Hesquiat Formation. Of these, *Zanthopsis vulgaris* is encountered the most frequently, not only on the west coast of Vancouver Island, but also in rocks of this age in Washington and Oregon. The oval carapace is marked by conspicuous furrows, including a major pair that isolates a vase-shaped central portion. The claws are massive—the right one about twice the size of the left. The complete specimen shown in Figure 152 was cracked out of a hard concretion in half a dozen pieces, and then reassembled with epoxy.

Figure 152 The crab *Zanthopsis vulgaris* (CDM 004), Hesquiat Fm., Carmanah Point.

Lyreidus (Figure 154). Recent species of *Lyreidus* live in deep water in the eastern Pacific Ocean between New Zealand and Japan, and in the Indian Ocean, where they burrow in mud. This raninid crab has a spindle-shaped smooth carapace with small lateral spines and a narrow front margin consisting of three spines. *Lyreidus* n. sp. from spherical concretions in the Via Appia Beds south of Campbell River differs from other species in bearing long lateral spines and a finely pitted carapace. This crab genus has a fossil record extending back 50 million years to the middle Paleogene. The Vancouver Island occurrence is from the initial part of this range.

Figure 153 The crab *Zanthopsis vulgaris* (RBCM), Hesquiat Fm., Carmanah Point and the bivalve *Chlamys* sp. (RBCM), Hesquiat Fm., Clo-ose.

Figure 154 The crab *Lyreidus* n. sp. (P. Bock Collection) and teredo-bored driftwood (VIPM 156). Both from Via Appia Beds, Appian Way site.

BARNACLES

Balanus (Figure 155). Charles Darwin wrote the definitive monographs on living and fossil barnacles in the early 1850s while contemplating his magnum opus, *The Origin of Species*. He characterized the present, not as the Age of Mammals, but as the Age of Barnacles. Those of us who live close to the shore around Vancouver Island or the Gulf Islands will probably concur with that statement. A barnacle, or cirriped, is a highly specialized crustacean that could be described as a shrimp-like creature cemented to a surface or substrate by its head, and that inhabits an armour of calcite plate. Its appendages, instead of being used for locomotion as in other crustaceans, are modified into a filter-feeding comb-like net. Intact fossil barnacles, such as *Balanus* sp., are uncommon in the Sooke Formation, but scattered porous plates are extremely common in the soft sandstones. These barnacle plates are triangular, with striate flanks, and because they are frequently stained black, these fossils are commonly mistaken for teeth.

MAMMALS

Fossils of two mammals are already known from the Sooke Formation on Vancouver Island—teeth of the desmostylian *Cornwallius* and a partial skull of the primitive whale *Chonecetus*. Desmostylians are an extinct group of large amphibious quadrupeds closely related to elephants and manatees. Their teeth consist of closely spaced enamel cylinders. These herbivores bore tusks on their lower jaws. Their hands and feet were in the form of paddles. They lived, hippopotamus-like, in and around the intertidal zone, where they fed on seaweed. *Chonecetus* belongs to the Archaeceti, a group of extinct whales that gave rise to both types of living whales—the baleen whales and the toothed whales.

Kolponomos (Figure 155). We now have evidence of a third type of fossil mammal in the Sooke Formation in the form of an incomplete lower jaw with an attached molar. This single fossil is similar to the jaw of the enigmatic *Kolponomos*—Greek for "the one that lives in a bay"—a large extinct carnivore that may be related to the raccoons and their allies, or to the sea lions and walruses. The bear-like *Kolponomos* probably prowled and swam along the shore searching for food, most likely mollusks and crustaceans, much like living sea otters.

Figure 155 The barnacle *Balanus* sp. (RBCM) and the lower jaw of the mammal *Kolponomos* (VIPM 145). Both Sooke Fm., near mouth of Muir Creek.

DICOT ANGIOSPERMS

Walnut Family (Figure 156). A concretionary layer near the base of the exposures of the Via Appia Beds contains a remarkable mixture of fossil specimens—commingled plant and animal, marine and nonmarine. Many different kinds of permineralized fruits and seeds become visible when these concretions are cut and polished. Some are so well preserved that the details of the interior structure is evident on the cellular level. A fruit of Juglandaceae, the walnut family, is a flattened nut with a pronounced keel at the equator. The symmetrical cotyledons, which become the primary seed leaves, are clearly evident as light chevrons separated by a septum.

Fertilized fruit (Figure 157). A unique permineralized specimen preserved in three dimensions from the Via Appia Beds shows the developing embryo within the fertilized fruit of an unidentified plant. Both the fruit and its attachment stalk are protected by sharp spines. The large white cells of the embryonic plant are clearly evident in the centre of the fruit.

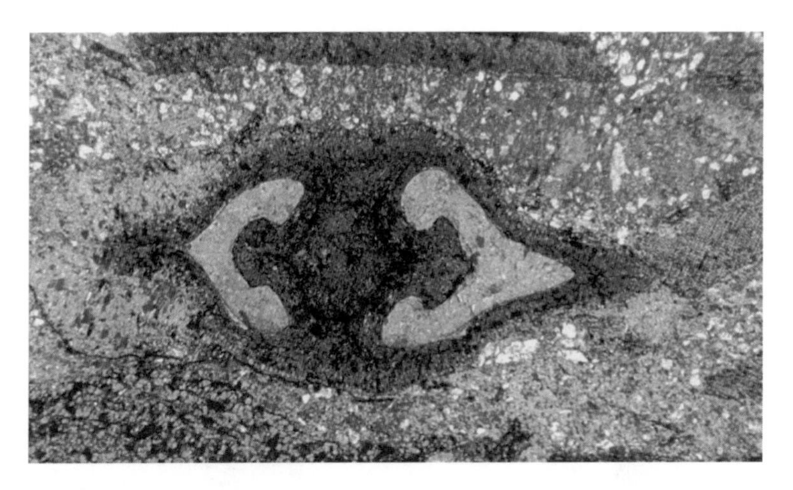

Figure 156 Cross-section of a fruit of the walnut family (VIPM 157). Fruit is 4 mm across. Via Appia Beds, Appian Way site.

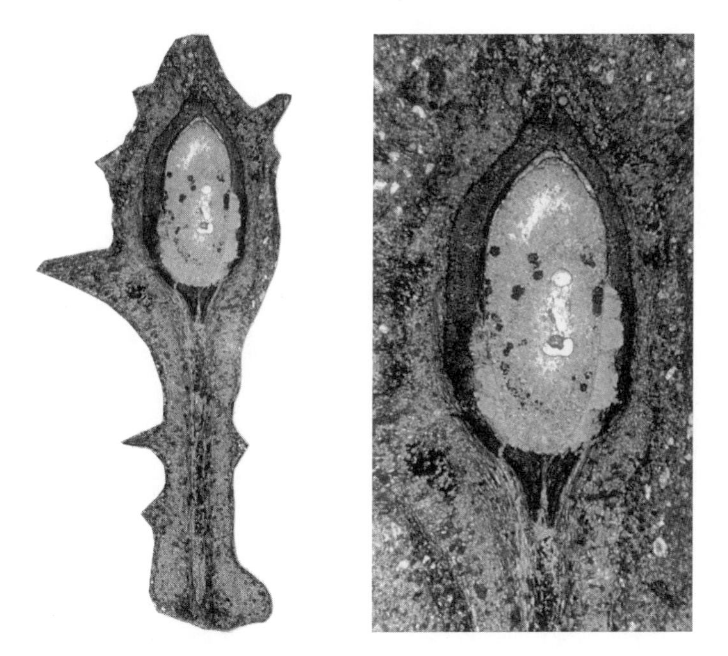

Figure 157 Fertilized fruit and attachment stem (6 mm long) and detail of fruit showing the developing embryonic cells (VIPM 158). Via Appia Beds, Appian Way site.

CHAPTER SEVEN

Collecting and Studying Fossils

Fossils may be found in all Phanerozoic sedimentary rocks. This does not mean that all shales, siltstones, sandstones, and limestones are fossiliferous—in fact, only a minority of such rocks contain well-preserved fossils. But it does mean that all of these rocks have the potential for yielding interesting and often important fossils. The search, however, is rarely easy. In addition to a bit of technical knowledge, it requires stamina and considerable care and patience.

If you are a novice fossil collector, we encourage you to accompany an experienced collector or a paleontologist on a couple of field trips before you attempt to extract fossils from rock by yourself. You might also want to join one of the regional paleontological societies on Vancouver Island or the mainland:

- Vancouver Island Paleontological Society, Courtenay
- Vancouver Island Paleontological Museum Society, Qualicum Beach

- Victoria Paleontology Society
- Vancouver Paleontological Society
- Thompson-Nicola Paleontological Society, Kamloops

Each of these societies has regular meetings where members learn about fossils and paleontology, and exchange information about Vancouver Island fossils. Regular field trips to different fossil localities are also organized. These five regional societies are members of the British Columbia Paleontological Alliance, an organization of amateur and professional paleontologists that strives to enhance knowledge about paleontology in the Cordillera and promote responsible collecting of fossils.

EQUIPMENT NEEDED

Only a few simple tools are necessary to safely extract most fossils from rock. The essential tools are a geological (or masonry) hammer and a 10-power hand lens, along with a few 10- to 16-cm cold chisels. A pair of safety goggles is mandatory to shield your eyes from flying rock chips or bits of steel. Leather work gloves protect your hands from cuts and scrapes, particularly if you are collecting barnacle-encrusted concretions from an intertidal area. A small crowbar is useful for prying out concretions, and a small sledge hammer may be required for tough rocks. Carry a good supply of newspapers to wrap large fossils and some small plastic vials for small fossils. Record each locality in a notebook and, before and after you extract them, be sure to describe how the fossils occur within the rock. This equipment and the fossil booty are best carried in a sturdy backpack.

PREPARING FOSSILS

Most fossil specimens require some preparation before they can be identified properly. In many cases, this involves nothing more than scrubbing them with a stiff plastic nail brush or an old toothbrush in water to remove any mud. If a fossil is partially enclosed in shale, test the rock before wetting it because some soft shales crumble when exposed to water. If the rock is hard, a concretion or limestone for example, a small chisel is useful for chipping away the matrix adhering to the fossil. Always have a tube of five-minute epoxy glue at hand to repair the fossils that fracture as a result of the preparation technique—something that is almost inevitable. Detailed work is best accomplished with a Vibro-Graver—a noisy miniature jackhammer—which should be used only under a binocular microscope. Do not apply shellac or clear nail polish to fossils because it is very difficult to photograph or study fossil specimens covered by these substances.

Those Permian and Triassic fossils on Vancouver Island that have been replaced by silica can be freed from the limestone matrix by submerging the blocks in weak hydrochloric acid. When handling such acid, always wear protective goggles and rubber gloves.

ORGANIZING A COLLECTION

Until a fossil collection has been identified, curated, and catalogued, it is merely a pile of peculiarly shaped rocks. Organizing a collection properly is a source of considerable personal satisfaction. As well, it is the best way to ensure that all identification and locality data are permanently recorded. Such data are mandatory if the fossils are ever to be used for paleontologic studies.

You should be able to identify most of the fossils collected from Vancouver Island rocks by reference to the photographs in this book. If a specimen cannot be identified to genus or species level, these photographs should at least indicate the larger group to which it belongs Then you should refer to the specialist publications and articles listed in the Reference section.

Each identified fossil should be assigned a unique catalogue number—your initials are an appropriate prefix to this number. Affix the number directly on the specimen with India ink. Record detailed locality information on an accompanying label along with the name of the formation. Other fossil specimens collected from the same site then receive the same locality number.

Below is an example of the layout and information included on a typical label to accompany each identified fossil in a well-organized collection—in this case, a Cretaceous heteromorph ammonite from shales exposed on the Puntledge River.

PAUL KRUGER FOSSIL COLLECTION

35 Baden Powell Street, Ladysmith, BC 339-9999

NAME:	*Bostrychoceras elongatum*
CATALOGUE NO.:	PK 305
LOCALITY NO.:	PR l 6
FORMATION:	Trent River Formation, lower part
LOCALITY:	Right side Puntledge River, 300 m downstream from Stotan Falls; concretion in grey shale
AGE:	Late Cretaceous
DESCRIPTION:	2 whorls of phragmocone, 5 cm across
DATE COLLECTED:	June 24, 1991

ETHICS OF COLLECTING

Unlike neighbouring Alberta, British Columbia has no legislation that limits or prohibits the collection of fossils by individuals. In Alberta, any fossil, be it a dinosaur bone, an ammonite, a bivalve, or a fossil leaf, cannot be collected, either by a professional or an amateur paleontologist, without a permit issued by the province. In British Columbia, by contrast, fossil collecting is specifically prohibited only in provincial and national parks, and in the few paleontological protected sites that have been established recently. So, while fossil collecting is implicitly permitted elsewhere, rules of common courtesy and safety prevail at all sites: do not collect on private property without permission of the owner, stay well away from machinery in active quarries, take care not to disturb oyster leases, remove any garbage, close gates, etc. In other words, as much as possible leave the area in the condition it was found.

According to the federal *Cultural Property Act*, fossils cannot be exported from Canada without a permit issued by Canadian Customs and Excise.

With regard to conservation and protection, fossil organisms are clearly not comparable to living organisms. Naturally, we would never advocate collecting the eggs of the marbled murrelet or digging up specimens of red-flowered gooseberry, both of which appear on endangered species lists. We do, however, suggest that a fragile ammonite should be carefully collected because, if this fossil remains in its natural rock setting, it may be destroyed by erosion or breakage. Simply put, the best place for a rare and important fossil is in a well-curated collection, preferably in a museum.

On the other hand, fossil collectors and amateur paleontologists should never attempt to remove permineralized bone of fossil reptiles or mammals. If you discover such a fossil, please contact a paleontologist at one of these institutions:

- Denman Institute for Research on Trilobites, Denman Island
- Vancouver Island Paleontological Museum, Qualicum Beach
- School of Earth and Ocean Sciences, University of Victoria
- Royal British Columbia Museum, Victoria
- Department of Geological Sciences, University of British Columbia
- Geological Survey of Canada, Vancouver

Fossils are irreplaceable records of past life. Most responsible collectors recognize that, although they may own a fossil in a strict sense, they are also stewards of a natural heritage that belongs to everyone. To this end, we strongly encourage fossil collectors to contribute locality information of identified fossil collections to provincial inventories so that paleontologists will know which fossils have been collected and who has collected the specimens of any particular fossil.

PHOTOGRAPHY OF FOSSILS

A good photograph, like a good painting, is a controlled distortion of reality. Some aspects of a subject are purposefully emphasized, while others are concealed. For the black and white photographs in this book, we prepared each fossil specimen so as to highlight its shape and surface texture, and to minimize its colour and surface gloss (see Figure 4). Consequently, these photographs appear somewhat different from actual specimens of fossils collected in the field. The procedure to achieve these images is outlined below.

First, carefully scrub the fossil and surrounding rock to remove all grease, dust, and mud. While it is still moist, blacken it with dilute black watercolour to produce an even, non-reflective, gun-metal-grey surface. Dry the specimen completely, preferably overnight. Immediately before photographing it, coat the fossil lightly with a sublimate of ammonium chloride—a salt that vapourizes in high heat. To do this, enclose the granular salt in a Pyrex glass tube with a nozzle at one end and a stopper connected to a flexible tube and a rubber bulb at the other end. Heat the tube over a Bunsen burner. As the salt vapourizes, spray the fumes across the blackened specimen by squeezing the bulb repeatedly The salt will coat the higher parts of the fossil, thus emphasizing the surface detail. With practice, this technique results in a light and evenly reflective coat on the fossil that can readily be photographed. Be sure to coat your fossil in a well-ventilated room (preferably in a fume hood) and, because ammonium chloride is hygroscopic (that is, it absorbs water from the atmosphere), avoid humid conditions.

We use a Rollei camera mounted vertically on a copying stand and professional-sized film (images are 6 cm square), but a regular 35 mm camera can also be used. Kodak Plus-X Pan film gives a good, fine-grained negative. To maximize the depth of field, use the smallest aperture possible (that is, the highest f-stop). Exposure times between 1/4 and 1 second generally produce good results. To avoid any confusion of convex and concave surfaces on a fossil, the main lighting direction should always be from the upper left. Place a white card on the lower right as a reflector to soften the shadows. In the darkroom, we use Ilford Multigrade paper with the highest filter (No. 7) to produce sharp, high-contrast prints.

After photography, remove the ammonium chloride coating from each fossil specimen by rinsing it under running water.

CHAPTER EIGHT

Fossil Localities on Vancouver Island

Fossils are known from literally hundreds of localities on Vancouver Island. Most of these sites are identified in a general way by the arrows on the geological maps (Figures 10, 22, 32, 42, 45–49, 146). Even if we had the space in this book to give directions to each and every locality, we wouldn't because the sense of personal discovery and serendipity provide such an enjoyable part of all natural history investigations. However, we do include information on how to find those localities that have produced many of the fossils illustrated in this book. Nine such localities are listed below—one from the Permian, one from the Triassic, one from the Jurassic, five from the Cretaceous, and one from the Cenozoic. With a single exception, these localities are readily accessible in that they are located less than a kilometre from a public road.

Serious fossil collectors should not confine their activities to these few localities. They are, after all, quite well known. Each exposure of shale, limestone, and sandstone on Vancouver Island and the Gulf Islands could potentially be an important fossil site.

PERMIAN

BUTTLE LAKE FORMATION AT MARBLE MEADOWS IN STRATHCONA PARK

Access to the fossiliferous Buttle Lake Formation at Marble Meadows in Strathcona Provincial Park is not easy. To reach this locality, you have to cross Buttle Lake and then hike up to Marble Meadows, a moderately strenuous endeavour. Take Highway 28 out of Campbell River to the Buttle Lake Bridge. Continue on the east side of the lake for 20 km to Augerpoint Picnic Site. Proceed by canoe 1 km straight across Buttle Lake to the mouth of Phillips Creek. Then follow the trail for 8 km to the A-frame shelter at Marble Meadows below Marble Peak. Continue past Marble Meadows along the height of land for another 3 km to the base of a steep north–south limestone face. At this locality, the Buttle Lake Formation consists of an enormous white and grey limestone mass entirely surrounded by Triassic volcanic rocks of the Karmutsen Formation. The white limestone is made up entirely of fossil fragments of crinoids along with poorly preserved silicified brachiopods. Well-preserved productid and spiriferid brachiopods and bryozoans are found in a 5-metre-thick unit of dark grey and thin-bedded limestone halfway up the ridge.

Remember that it is illegal to collect fossils in a provincial park without a permit issued by BC Parks.

TRIASSIC

QUATSINO FORMATION AT OPEN BAY ON QUADRA ISLAND

The ferry from Campbell River docks at Quathiaski Cove on Quadra Island. From here drive north on West Road (which becomes Hyacinthe Bay Road) for about 18 km. After passing September Lake, turn south onto Valdez Road. Drive for another 6 km and turn right on

Breton Road, or else continue along Valdez Road. Either road terminates at a public access to the shoreline. Along the shore, exposed limestones of the Quatsino Formation are crinkled into tight folds outlined by hard chert bands. The fossils here are primarily small ammonites that weather-out from the limestones because they have been replaced by silica. Some of the ammonites are deformed and stretched beyond recognition; others are very well preserved.

JURASSIC

BONANZA GROUP ON THE MEMEKAY RIVER

The instructions of how to get to this fossil site are rather complex. If you are driving north on the Island Highway, turn left onto Sayward Road at the Sayward Junction. Then turn right onto Hern Road. At the end of Hern Road take the upper left road (called C-Branch). Set your car's trip odometer to 0 here. At 18.5 km cross a bridge. At 29.8 km leave C-Branch and turn left onto Memekay Main. At 30.5 km cross another bridge. Take the first right after the bridge and, at 31.8 km, turn left and park. The blocks of fossiliferous siltstones of the Bonanza Group (Lower Jurassic) are found about 100 m further, on the left side of the road. Look for the large evolute ammonite *Paltechioceras*, bivalves, and gastropods.

Because the fossiliferous exposures of the Bonanza Group are located along a busy logging road in an active logging area in the Memekay Valley, please do not attempt to visit this site on a working day.

CRETACEOUS

TRENT RIVER FORMATION ON THE PUNTLEDGE AND BROWNS RIVERS

From downtown Courtenay, take Lake Trail Road west for about 2 km and turn right on the Comox Logging Road. Keep on this road past the Fletcher Challenge works yard for another 2 km until reaching the high bridge that crosses the Puntledge River at Stotan Falls. Park here and take the trail down to the river. A broad expanse of fossiliferous siltstone of the lower Trent River Formation is exposed at and above the falls. The large ribbed ammonite *Eupachydiscus* is a characteristic fossil here. The Browns River joins the Puntledge about 1 km downstream from the falls. Fossiliferous shales are exposed discontinuously along both the Puntledge and Browns rivers. Bivalves, gastropods, ammonites, crustaceans, and fish remains are fairly common fossils. Bones of marine reptiles such as mosasaurs, elasmosaurs, and turtles have also been found. A 50-cm-thick sandstone bed at the confluence of the Browns and Puntledge contains abundant fossil leaves. Collecting is not permitted at the Puntledge Elasmosaur Site on the south side of the Puntledge, about 200 m downstream from the confluence of the Browns.

TRENT RIVER FORMATION AT NORTHWEST BAY

Northwest Bay is located just south of Parksville. Turn east off the Island Highway onto Northwest Bay Road. Continue for 3 km and then turn left onto Wall Beach Road, which ends in a parking area up a short hill. Take the trail to the beach. The first beds encountered are yellow-brown sandstones with trigoniid bivalves. Overlying these beds are fossiliferous, gritty, blue-grey shales with bivalves, gastropods, ammonites, and crustaceans.

HASLAM FORMATION AT BRANNAN LAKE

Brannan Lake is located on the flank of Mount Benson, just west of Nanaimo. Turn off the Island Highway at Hammond Bay Road, continue for about 100 m and then turn left onto Metral Drive. Turn right onto Dumont Road and continue for about 3 km. The road remains blacktop right up to two abandoned shale pits. The lower pit is presently leased by the Nanaimo Motorcycle Club and should be entered only with permission and when races are not taking place. Abundant ammonites, bivalves, gastropods, and fish remains can be found in the shales of the Haslam Formation and in the concretions. The ammonite *Canadoceras* is common at this locality and, because of its chalky white shell, it is a conspicuous fossil. At this locality, the concretions frequently contain fish coprolites and shark teeth.

OYSTER BAY FORMATION AT SHELTER POINT

Shelter Point is located just off the Island Highway, about 10 km south of downtown Campbell River. At the northern end of Shelter Bay, turn east onto Heard Road, which ends at a public access to Shelter Point. A low tide is necessary in order to collect from these shales. The fossils, mainly the crab *Longusorbis* and the straight ammonite *Baculites*, occur only in the gritty concretions that weather-out of the shale.

LAMBERT FORMATION AT COLLISHAW POINT ON HORNBY ISLAND

The ferry from Denman Island docks at Shingle Spit on Hornby Island. From here follow Shingle Spit Road until, at the sharp turn, it becomes Central Road. After 500 m turn left onto Savoie Road and drive north and park at the turnaround. A trail to the beach follows a public access on the left side of the fence. On reaching the beach, turn right to face Collishaw Point—known locally as Boulder Point on account of the large number of rounded granite boulders. A low tide, the lower the better, is necessary to collect from this well-frequented site. Fossils occur in the shales, but they are particularly common and well preserved in the rounded concretions. Ammonites and bivalves (mainly *Pachydiscus*, *Baculites*, and *Inoceramus*), which retain the original mother-of-pearl shell layer, are characteristic fossils at this site. The spectacular heteromorph *Nostoceras* occurs only at this locality The shale cliffs at the base of Collishaw Point are also very fossiliferous.

CENOZOIC

SOOKE FORMATION AT MUIR CREEK

From the town of Sooke west of Victoria, follow Highway 14 for about 14 km to the Muir Creek Campground. Park here and walk north along the beach. After 400 m, low cliffs of soft, pebbly sandstones of the Sooke Formation begin. Abundant fossils are evident in the cliffs and in the yellow-brown sandstones exposed below the high tide level on the beach. Bivalves, gastropods, and barnacle plates are extremely common in the soft friable sandstone, which also contains a few plant remains, as well as rare teeth and bones of marine mammals.

Glossary

Ammonite An extinct cephalopod with a coiled shell, complex sutures, and a siphuncle located along the outer margin.
Angiosperms Flowering plants that protect their seeds within an ovary.
Aptychus An ammonite jaw in the form of a wrinkled, folded, calcareous plate.
Basalt A fine-grained and dark-coloured volcanic rock.
Belemnite An extinct squid-like cephalopod with a solid, bullet-shaped internal shell.
Biostratigraphy The study of the succession of fossil animals and plants in bedded rocks and their significance.
Body fossil The actual remains of an organism preserved in rock.
Calcareous Composed of calcite (calcium carbonate).
Chert A fine-grained glassy rock composed of silica.
Coast Belt The belt of igneous rock along the Pacific coast formed from the collision of the Insular and Intermontane belts.
Comox Basin The Upper Cretaceous depositional basin now located on east-central Vancouver Island and Denman and Hornby islands.
Concretion A hard calcareous nodule formed in softer sedimentary rock; occasionally around a fossil.
Conglomerate A sedimentary rock composed of gravel-sized, rounded particles, surrounded by sand.
Conodonts Phosphatic tooth-like microfossils; the remains of extinct jawless fishes.
Continental drift The study of the shift of continents relative to each other across the surface of the Earth.
Coprolite Fossilized animal excrement.
Corallite An individual of a colonial coral.
Cordillera The series of mountain ranges and chains located along the west coast of the Americas.
Cryptozoic The first eon of geologic time, which lasted for more than 3,000 million years. It contains only simple, unicellular fossil organisms.
Decapoda Crustaceans, such as crabs and lobsters, that bear five pairs of appendages on the thorax.
Desmostylian A large extinct amphibious quadruped related to elephants and manatees.
Dicots Angiosperms with floral parts in fours or fives and net-veined leaves, such as broad-leafed trees, shrubs, and flowers.
Divaricate A type of surface sculpture of bivalves consisting of chevron-shaped ribs crossing the growth lines.
Dorsal Referring to the upper side of an animal.
Eon The largest division of geologic

time. Two eons are recognized: Cryptozoic and Phanerozoic.

Epifaunal Animals, such as snails, crabs, and oysters, that live on top of the sediment.

Era A division of geologic time shorter than an eon. Five eras are recognized: Archean, Proterozoic, Paleozoic, Mesozoic, and Cenozoic.

Eucaryotic cell An advanced cell type having a nucleus containing chromosomes that is found in all organic kingdoms except Monera.

Evolute Coiling of ammonites and nautiloids with all the whorls visible.

Foreland Belt The belt of faulted sedimentary rock forming the western margin of Laurentia.

Formation A distinctive body of sedimentary or volcanic rock that forms a convenient stratigraphic unit.

Fossil The remains, traces, or impressions of ancient animals and plants naturally preserved in sedimentary rocks.

Fungi A kingdom of complex multicellular organisms; plant-like, but requiring external food sources—that is, moulds and mushrooms.

Gastroliths Stones swallowed by vertebrate animals to aid in grinding food and, possibly, to serve as ballast.

Geochronology The study of the numerical ages of sedimentary and igneous rocks.

Group A body of sedimentary or volcanic rock comprising two or more formations.

Gymnosperms Plants, such as conifers, cycads, and maidenhair trees, that bear exposed seeds.

Half-life The time required for half of a parent isotope to decay to a daughter isotope.

Heteromorph An ammonite that is not planispirally coiled.

Igneous A rock formed by cooling of molten magma.

Infaunal Animals, such as clams, that live within the sediment.

Inoceramid A bivalve of the family Inoceramidae having prominent concentric undulations to the shell and lacking teeth.

Insular Belt The area of volcanic and sedimentary rock along the Pacific coast composed of amalgamated Alexander Terrane and Wrangellia.

Intermontane Belt The area of volcanic and sedimentary rock in central BC composed of amalgamated Cache Creek Terrane, Quesnellia, and Stikinia.

Involute Coiling in ammonites and nautiloids with the last whorl covering previously formed whorls.

Isotope One of two varieties of an element that differ only in the number of neutrons in the nucleus.

Kingdom The highest category of organisms. Five kingdoms are recognized on the basis of cell type, cellular organization, and means of acquiring food.

Laurentia The ancient continent that was the forerunner to North America. Only the eastern third of British Columbia is part of Laurentia.

Lira Fine, thread-like, raised lines on the outside of a shell.

Lophophore The respiration and food-gathering organ of a brachiopod.

Magma Molten rock.
Metamorphic A rock that has been altered and recrystallized by heat and presssure.
Metaphyta A kingdom of complex, eucaryotic, multicellular organisms that manufacture their own food—that is, plants.
Metazoa A kingdom of complex, eucaryotic, multicellular organisms that depend on external food sources—that is, animals.
Monera A kingdom of minute, procaryotic, unicellular organisms, such as bacteria and blue-green algae.
Monocots Angiosperms, such as palms and grasses, that have floral parts in threes and parallel-veined leaves.
Nanaimo Basin The Upper Cretaceous depositional basin now located on southeastern Vancouver Island, the southern Gulf Islands, and part of the San Juan Islands.
Nautiloid A cephalopod with a coiled shell, simple sutures, and a central siphuncle.
Omineca Belt The belt of igneous and metamorphic rock in eastern BC formed from the collision of the Intermontane and Foreland belts.
Oxycone An involute shell of a compressed ammonite with a sharp venter.
Paleontology The study of ancient life as represented by fossils preserved in rocks.
Pangaea The supercontinent that assembled in the Permian and rifted apart in the Middle Jurassic.
Permineralization A type of fossil preservation in which the porous spaces in bone and wood are infilled by the minerals calcite or quartz.
Phanerozoic The second eon of geologic time, which lasted for the last 650 million years. It contains fossils of complex, multicellular organisms.
Phragmocone The part of the shell of ammonites or nautiloids that is divided by septa into a series of gas-filled chambers.
Phylum A category of organisms that share a major type of body plan. There are about twenty-five living phyla in the kingdom Metazoa (animals)—for example, Brachiopoda, Echinodermata, Arthropoda, Mollusca, Chordata.
Planispiral coiling Bilaterally symmetrical coiling in one plane, characteristic of most ammonites and nautiloids.
Plate tectonics The study of the movement and interaction of rigid plates of oceanic and continental crust resulting in the formation of large-scale geologic features, such as mountains.
Procaryotic cell A primitive cell type lacking a nucleus that is restricted to the kingdom Monera.
Protoctista A kingdom of eucaryotic, unicellular organisms, such as ameboids and protists.
Pseudofossil Inorganically produced stony objects that mimic real fossils.
Radula An elongate row of horny or mineralized teeth in the mouth region of most mollusks.
Raninid A crab of the family Raninidae, characterized by a shield-shaped carapace, puny claws, and a flexible abdomen.

Rodinia The ancient supercontinent that rifted apart in the latest Cryptozoic.
Sandstone A sedimentary rock composed of sand particles of quartz.
Sea-floor spreading The study of the formation and destruction of oceanic crust.
Sedimentary A rock formed by the accumulation of weathered particles of existing rock or of organic particles.
Sepiid A cephalopod of the order Sepiida (cuttlefish) that has a porous internal shell.
Serpenticone The evolute shell of an ammonite that coils like a rope.
Shale A sedimentary rock composed of clay-sized particles.
Siphuncle A porous tube connecting all the chambers of nautiloids and ammonites.
Stratigraphy The study of the relationship and succession of bedded sedimentary and volcanic rocks (strata).
Sutures The line of intersection of a septum and the outer wall in nautiloids and ammonites.
System A division of geologic time shorter than an era. The Phanerozoic Eon comprises thirteen systems—Vendian to Quaternary.
Terrane An area of the Cordillera characterized by distinctive rock types indicating a geologic history different from other areas.
Terrane tectonics The study of the accretion of terranes along the margin of a continent to form broad mountain belts, such as the Cordillera.
Theropod A bipedal carnivorous dinosaur.
Trace fossil The tracks, trails, or burrows of animals preserved in sedimentary rock.
Trigoniid A bivalve of the family Trigoniidae, characterized by triangular, strongly ornamented valves and large, deeply fluted teeth.
Trilobite An extinct arthropod with three longtitudinal lobes.
Tuff A sedimentary rock composed of volcanic particles.
Umbo The first-formed part of a valve.
Valve One of the distinct elements of which a shell is made. Some organisms have a single valve (limpet); others have hinged paired valves (bivalves, brachiopods).
Venter The outside edge of a coiled ammonite or nautiloid.
Ventral Referring to the lower side of an animal.
Wrangellia The predominantly volcanic terrane that now forms Vancouver Island, Haida Gwaii (the Queen Charlotte Islands), and parts of Alaska.

References

The books and scientific journals listed below are specialized geological or paleontological publications that generally are not available in regular bookstores or public libraries. They can be examined in university libraries. Geological Survey of Canada publications can be purchased at the West Coast office at #101, 605 Robson Street, Vancouver, BC V6B 5J3.

HISTORICAL GEOLOGY AND PALEONTOLOGY

Albritton, C.C. Jr. 1980. *The Abyss of Time: Changing Concepts of the Earth's Antiquity after the Sixteenth Century*. J.P. Tarcher, Inc.

Boardman, R.S., Cheetham, A.H., and Rowell, A.J., eds. 1987. *Fossil Invertebrates*. Blackwell Scientific Publications.

Bromley, R.G. 1990. *Trace Fossils: Biology and Taphonomy*. Unwin Hyman.

Carroll, R.L. 1988. *Vertebrate Paleontology and Evolution*. W.H. Freeman and Co.

Clarkson, E.N.K. 1986. *Invertebrate Palaeontology and Evolution*, Second Edition. Allen & Unwin.

Lehmann, U. 1981. *The Ammonites: Their Life and their World*. Cambridge University Press.

Ludvigsen, R., ed. 1996. *Life in Stone: A Natural History of British Columbia's Fossils*. University of British Columbia Press.

MacFall, R.P., and Wollin, J.C. 1972. *Fossils for Amateurs: A Guide to Collecting and Preparing Invertebrate Fossils*. Van Nostrand Reinhold Company.

McKerrow, W.S., ed. 1978. *The Ecology of Fossils*. The MIT Press.

Murray, J.W., ed. 1985. *Atlas of Invertebrate Macrofossils*. John Wiley & Sons.

Raup, D.M., and Stanley, S.M. 1978. *Principles of Paleontology*, Second Edition. W.H. Freeman and Company.

Reader, J. 1986. *The Rise of Life: The First 3.5 Billion Years*. Alfred A. Knopf.

Rudwick, M.J.S. 1972. *The Meaning of Fossils: Episodes in the History of Palaeontology*. American Elsevier Inc.

Stanley, S.M. 1988. *Earth and Life through Time*. W.H. Freeman and Company.

Stearn, C.W., Carroll, R.L., and Clark, T.H. 1979. *Geological Evolution of North America*. John Wiley & Sons.

Tidwell, W.D. 1975. *Common Fossil Plants of Western North America*. Brigham Young University Press.

CORDILLERAN GEOLOGY

Gabrielse, H., and Yorath, C.J., eds. 1991. *Geology of the Cordilleran Orogen in Canada*. Geological Survey of Canada, Geology of Canada, No. 4.

Jones, D.L., Cox, A., Coney, P., and Beck, M. 1989. "The growth of western North America." *Scientific American*, vol. 264, pp. 70–84.
Monger, J.W.H. 1993. "Canadian Cordilleran tectonics: Geosynclines to crustal collage." *Canadian Journal of Earth Sciences*, vol. 30, pp. 209–31.
Yorath, C.J. 1990. *Where Terranes Collide*. Orca Book Publishers.

VANCOUVER ISLAND ROCKS

Buckham, A.F. 1947. "The Nanaimo Coal Field." *Transactions of the Canadian Institute of Mining and Metallurgy*, vol. 50, pp. 460–72.
Cameron, B.E.B. 1980. *Biostratigraphy and depositional environment of the Escalante and Hesquiat formations (Early Tertiary) of the Nootka Sound Area, Vancouver Island, British Columbia*. Geological Survey of Canada, Paper 78-9.
Carlisle, D., and Susuki, T. 1965. "Structure, stratigraphy and paleontology of an Upper Triassic section on the West Coast of British Columbia." *Canadian Journal of Earth Sciences*, vol. 2, pp. 442–84.
Carlisle, D., and Susuki, T. 1974. "Emergent basalt and submergent carbonate-clastic sequences including the Upper Triassic Dilleri and Welleri zones on Vancouver Island." *Canadian Journal of Earth Sciences*, vol. 11, pp. 254–79.
Carson, D.J.T. 1973. *The plutonic rocks of Vancouver Island*. Geological Survey of Canada, Paper 72-44.
Clapp, C.H. 1912. *Southern Vancouver Island, British Columbia*. Geological Survey of Canada, Memoir 13.
Clapp, C.H. 1914. *Geology of the Nanaimo Map-area*. Geological Survey of Canada, Memoir 51.
Clapp, C.H. 1917. *Sooke and Duncan Map-areas, Vancouver Island*. Geological Survey of Canada, Memoir 96.
Clapp, C.H., and Shimer, H.W. 1911. "The Sutton Jurassic of the Vancouver Group, Vancouver Island." *Proceedings of the Boston Society of Natural History*, vol. 134, pp. 425–38.
England, T.D.J. 1989. *Lithostratigraphy of the Nanaimo Group, Georgia Basin, southwestern British Columbia*. Geological Survey of Canada, Paper 89-1E, pp. 197–206.
England, T.D.J., and Calon, T.J. 1991. "The Cowichan fold and thrust system, Vancouver Island, southwestern British Columbia." *Geological Society of America Bulletin*, vol. 103, pp. 336–62.
England, T.D.J., and Hiscott, R.N. 1992. "Lithostratigraphy and deep-water setting of the Upper Nanaimo Group (Upper Cretaceous), outer Gulf Islands of southwestern British Columbia." *Canadian Journal of Earth Sciences*, vol. 29, pp. 574–95.
Fyles, J.T. 1955. *Geology of the Cowichan Lake area, Vancouver Island, British Columbia*. BC Department of Mines, Bulletin 37.
Jeletzky, J.A. 1975. *Hesquiat Formation (new): A neritic channel and inter-channel deposit of Oligocene age, western Vancouver Island, British Columbia*. Geological Survey of Canada, Paper 75-32.
Jeletzky, J.A. 1976. *Mesozoic and Tertiary rocks of Quatsino Sound, Vancouver Island, British Columbia*. Geological Survey of Canada, Bulletin 242.

Jones, D.L., Silberling, N.J., and Hillhouse, J. 1977. "Wrangellia—a displaced terrane in northwestern North America." *Canadian Journal of Earth Sciences*, vol. 14, pp. 2565–77.

Massey, N.W.D., and Friday, S.J. 1987. "Geology of the Chemainus River–Duncan area, Vancouver Island." *Geological Fieldwork*, BC Ministry of Energy, Mines and Petroleum Resources, pp. 81–91.

McLellan, R. 1927. "Geology of the San Juan Islands." *University of Washington Publications in Geology*, No. 2.

Muller, J.E. 1980. *The Paleozoic Sicker Group of Vancouver Island, British Columbia*. Geological Survey of Canada, Paper 79-30.

Muller, J.E., Cameron, B.E.B., and Northcote, K.E. 1981. *Geology and mineral deposits of Nootka Sound Map-area, Vancouver Island, British Columbia*. Geological Survey of Canada, Paper 80-16.

Muller, J.E., and Jeletzky, J.A. 1970. *Geology of the Upper Cretaceous Nanaimo Group, Vancouver Island and Gulf Islands, British Columbia*. Geological Survey of Canada, Paper 69-25.

Mustard, P.S. 1994. *The Upper Cretaceous Nanaimo Group, Georgia Basin*. Geological Survey of Canada Bulletin 481, pp. 27-95.

Mustard, P.S., and Rouse, G.E. 1994. *Stratigraphy and evolution of Tertiary Georgia Basin and subjacent Upper Cretaceous sedimentary rocks, southwestern British Columbia and northwestern Washington State*. Geological Survey of Canada Bulletin 481, pp. 97-169.

Roddick, J.A., Muller, J.E., and Okulitch, A.V. 1979. *Fraser River, British Columbia–Washington*. Geological Survey of Canada, Map 1386A.

Yole, R.W. 1969. "Upper Paleozoic stratigraphy of Vancouver Island, British Columbia." *Proceedings of the Geological Association of Canada*, vol. 20, pp. 30–40.

Yole, R.W., and Irving, E. 1980. "Displacement of Vancouver Island: Paleomagnetic evidence from the Karmutsen Formation." *Canadian Journal of Earth Sciences*, vol. 17, pp. 1210–28.

VANCOUVER ISLAND FOSSILS

Anderson, F.M. 1958. *Upper Cretaceous of the Pacific Coast*. Geological Society of America, Memoir 71.

Bell, W.A. 1957. *Flora of the Upper Cretaceous Nanaimo Group of Vancouver Island, British Columbia*. Geological Survey of Canada, Memoir 293.

Clark, B.L., and Arnold, R. 1923. "Fauna of the Sooke Formation, Vancouver Island." *University of California Publications in Geological Sciences*, vol. 14, pp. 123–234.

Feldmann, R.M., and McPherson, C.B. 1980. *Fossil decapod crustaceans of Canada*. Geological Survey of Canada, Paper 79-16.

Frebold, H. 1964. *Illustrations of Canadian fossils: Jurassic of western and Arctic Canada*. Geological Survey of Canada, Paper 63-4.

Frebold, H., and Tipper, H.W. 1970. "Status of the Jurassic in the Canadian Cordillera of British Columbia, Alberta and southern Yukon." *Canadian Journal of Earth Sciences*, vol. 7, pp. 1–21.

Fritz, M.A. 1932. "Permian Bryozoa from Vancouver Island." *Transactions of the Royal Society of Canada*, 3rd series, vol. 29, pp. 149–61.

Haggart, J.W. 1989. *New and revised ammonites from the Upper Cretaceous Nanaimo Group of British Columbia and Washington State*. Geological Survey of Canada, Bulletin 396, pp. 181–221.

Haggart, J.W. 1991. A *new assessment of the age of the basal Nanaimo Group, Gulf Islands, British Columbia*. Geological Survey of Canada, Paper 91-1E, pp. 77-82.

Haggart, J.W. 1991. "Biostratigraphy of the Upper Cretaceous Nanaimo Group, Gulf Islands, British Columbia," in Smith, P.L., ed., A *Field Guide to the Paleontology of Southwestern Canada*, Geological Association of Canada, pp. 223–56.

Haggart, J.W. 1996. "Mollusks: Exotic shells from Cretaceous seas," in Ludvigsen, R., ed., *Life in Stone: A Natural History of British Columbia's Fossils*, UBC Press, pp. 167-86.

Haggart, J.W., and Ward, P.D. 1989. "New Nanaimo Group ammonites (Cretaceous, Santonian–Campanian) from British Columbia and Washington State." *Journal of Paleontology*, vol. 63, pp. 218–27.

Jeletzky, J.A. 1965. *Late Upper Jurassic and Early Lower Cretaceous fossil zones of the Canadian Western Cordillera, British Columbia*. Geological Survey of Canada, Bulletin 103.

Jones, D.L. 1963. *Upper Cretaceous (Campanian and Maastrichtian) ammonites from southern Alaska*. US Geological Survey Professional Paper 432.

Ludvigsen, R. 1996. "Ancient saurians: Cretaceous reptiles of Vancouver Island." in Ludvigsen, R., ed., *Life in Stone: A Natural History of British Columbia's Fossils*, UBC Press, pp. 156–66.

McKenzie McAnally, L. 1996. "Paleogene mammals on land and at sea," in Ludvigsen, R., ed., *Life in Stone: A Natural History of British Columbia's Fossils*, UBC Press, pp. 202–11.

Meek, F.B. 1858. "Descriptions of new organic remains from the Cretaceous rocks of Vancouver's Island." *Transactions of the Albany Institute*, vol. 4, pp. 37–49.

Merriam, J.C. 1899. *The fauna of the Sooke Beds of Vancouver Island*. Proceedings of the California Academy of Sciences, Geology, vol. 1, pp. 175–80.

Nicholls, E.L. 1992. "Note on the occurrence of the marine turtle *Desmatochelys* (Reptilia: Chelonioidea) from the Upper Cretaceous of Vancouver Island." *Canadian Journal of Earth Sciences*, vol. 29, pp. 377–80.

Rathburn, M.J. 1926. *The fossil stalk-eyed Crustacea of the Pacific slope of North America*. United States National Museum, Bulletin 138.

Richards, B.C. 1975. "*Longusorbis cuniculosus*: A new genus and species of Upper Cretaceous crab; with comments on the Spray Formation at Shelter Point, Vancouver Island, British Columbia." *Canadian Journal of Earth Sciences*, vol. 12, pp. 1850–63.

Russell, L.S. 1968. "A new cetacean from the Oligocene Sooke Formation of Vancouver Island, British Columbia." *Canadian Journal of Earth Sciences*, vol. 5, pp. 929–33.

Saul, L.R. 1978. "The North Pacific Cretaceous trigoniid genus *Yaadia*." *University of California Publications in Geological Sciences*, vol. 119.
Smith, P.L., and Tipper, H.W. 1986. "Plate tectonics and paleobiogeography: Early Jurassic (Pleinsbachian) endemism and diversity." *Palaios*, vol. 1, pp. 399–412.
Tozer, E.T. 1967. *A standard for Triassic time*. Geological Survey of Canada, Bulletin 156.
Tozer, E.T. 1979. *Latest Triassic ammonoid faunas and biochronology, Western Canada*. Geological Survey of Canada, Paper 79-1B, pp. 127–35.
Tozer, E.T. 1984. *The Trias and its ammonoids: The evolution of a time scale*. Geological Survey of Canada, Miscellaneous Report 35.
Usher, J.L. 1952. *Ammonite faunas of the Upper Cretaceous rocks of Vancouver Island*. Geological Survey of Canada, Bulletin 21.
Waldman, M. 1971. "Hexanchid and orthacondontid shark teeth from the Lower Tertiary of Vancouver Island." *Canadian Journal of Earth Sciences*, vol. 8, pp. 166–70.
Ward, P.D. 1976. "Upper Cretaceous ammonites (Santonian–Campanian) from Orcas Island, Washington." *Journal of Paleontology*, vol. 50, pp. 454–61.
Ward, P.D. 1978. "Revisions to the stratigraphy and biochronology of the Upper Cretaceous Nanaimo Group, British Columbia and Washington State." *Canadian Journal of Earth Sciences*, vol. 15, pp. 405–23.
Ward, P.D. 1978. "Baculitids from the Santonian–Maastrichtian Nanaimo Group, British Columbia, Canada and Washington State, USA." *Journal of Paleontology*, vol. 52, pp. 1143–54.
Ward, P.D. 1985. "Upper Cretaceous (Santonian–Maastrichtian) molluscan faunal associations, British Columbia," in Bayer, U., and Seilacher, A., eds., *Sedimentary and Evolutionary Cycles*, Springer-Verlag, pp. 397–420.
Ward, P.D., and Mallory, V.S. 1977. "Taxonomy and evolution of the lytoceratid genus *Pseudoxybeloceras* and relationship to the genus *Solenoceras*." *Journal of Paleontology*, vol. 51, pp. 6606–18.
Whiteaves, J.F. 1879. "On the fossils of the Cretaceous rocks of Vancouver and adjacent islands in the Strait of Georgia." *Geological Survey of Canada, Mesozoic Fossils*, vol. 1, pp. 93–190.
Whiteaves, J.F. 1895. "On some fossils from the Nanaimo Group of the Vancouver Cretaceous." *Transactions of the Royal Society of Canada*, 2nd ser., vol. 1, pp. 119–33.
Whiteaves, J.F. 1903. "On some additional fossils from the Vancouver Cretaceous, with a revised list of species therefrom." *Geological Survey of Canada, Mesozoic Fossils*, vol. 1, pp. 309–416.
Yole, R.W. 1963. "An Early Permian fauna from Vancouver Island, British Columbia." *Bulletin of Canadian Petroleum Geology*, vol. 11, pp. 138–49.

Index

The following abbreviations are used in the index:

(am) ammonite	(cu) cuttlefish	(na) nautiloid
(be) belemnite	(di) dinosaur	(pl) plant
(bi) bivalve	(ec) echinoderrn	(re) reptile
(br) brachiopod	(fi) fish	(sc) scaphopod
(cd) conodont	(ga) gastropod	(tr) trace fossil
(co) coral	(in) insect	(tb) trilobite
(cr) crustacean	(ma) mammal	